Troubleshooting With the Oscilloscope

by
Robert G. Middleton

Howard W. Sams & Co., Inc.
4300 WEST 62ND ST. INDIANAPOLIS, INDIANA 46268 USA

Preface

When the third edition of this book was published in 1975, trouble-shooting with an oscilloscope had become widely accepted as the most reliable method for analysis of malfunctions in electronic circuitry. With the rapid advance of electronic technology the oscilloscope has gained universal acceptance as an indispensable test and measurement instrument. This book is intended to help you obtain the maximum benefits from an oscilloscope and to "painlessly" familiarize you with the wide selection of oscilloscopes now available to service technicians. It was planned and written with a full appreciation of the type of practical instruction that technicians need.

Not all electronics troubleshooters will be concerned with the most sophisticated types of oscilloscope applications and displays. Therefore this book is directed toward obtaining the maximum benefits from an oscilloscope, even if you have never used the instrument before. To begin with, you will learn the basic oscilloscope waveforms in Chapter 1. Peak-to-peak voltage measurements, as specified in standard service data, are described and illustrated. The use of triggered sweep in detection of "hidden" distortion is described and illustrated. Since most tv waveforms are various types of complex exponentials, the natural law of growth and decay is discussed, with relevant examples. Oscilloscope test classifications—signal tracing, waveform checks, frequency response, phase checks, frequency measurements, time measurements, and transient tests—are reviewed. Because there is a definite trend to the use of "intelligent" (or "smart") oscilloscopes, this topic has been included. Another new oscilloscope for service applications provides dual-channel display, with ("smart") A and B, A + B, and A − B modes.

In the second chapter a detailed discussion of oscilloscope operation is provided. Both free-running and triggered sweeps are explained; single-trace and dual-trace operations are described and illustrated; intensity, position, focus, astigmatism, sync, time base, gain, and triggered-sweep controls are covered. Chapter 3 provides practical information concerning the applications and functions of

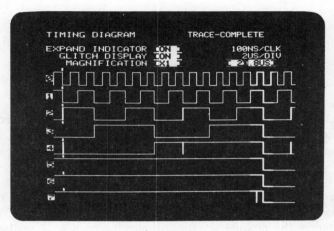

(A) Time/frequency pattern.

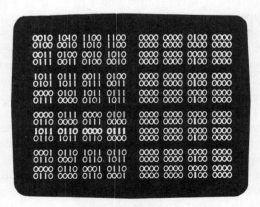

(B) Data-domain pattern.

Courtesy Hewlett-Packard, Inc.

Fig. 1. Two basic types of digital displays.

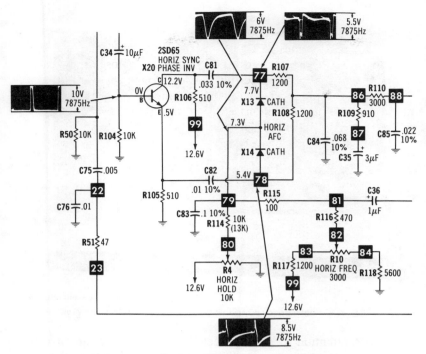

Fig. 2. Example of specified peak-to-peak voltages.

various oscilloscope probes. It is explained why probes are needed, how low-capacitance probes are constructed and adjusted, how demodulator probes are constructed and how they operate, and how a resistive "isolating" probe functions. Although high-voltage capacitance-divider probes are used more extensively in electronic labs than in service shops, the construction and operation of this type of probe have been included. Notes are provided on suitable kinds of generators in various classes of probe tests.

The fourth chapter presents a practical discussion of signal-tracing techniques and the evaluation of oscilloscope displays. Troubleshooting of rf amplifiers, if amplifiers, and video amplifiers, with identification of distortion factors, is described and illustrated. Tuned-circuit frequency response and visual-alignment procedures are noted. Common difficulties are covered, and device malfunction is discussed briefly. Chapter 5 proceeds into details of sync-section signal tracing. Oscilloscope application techniques, sync-channel bandwidth, and sync-inverter action are presented. Color-sync troubleshooting is cov-

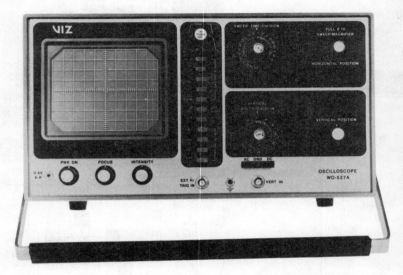

Fig. 3. A triggered-sweep oscilloscope with a tv time selector control.

ered, with attention to phase-detector and color-sync circuit action. The dual-time-constant sync clipper is noted, and sync-section troubleshooting in older models of receivers is reviewed. Information is included on modular construction.

Chapter 6 discusses troubleshooting the afc and horizontal-oscillator section, and common causes of malfunction are explained. Symptoms of device defects are noted. Since the tv technician must still cope to an appreciable extent with tube-type receivers, troubleshooting in tube-type afc and horizontal-oscillator circuitry is reviewed. Chapter 7 treats waveform tests in the horizontal-sweep section, with particular attention being given to the widely used SCR horizontal-sweep system. The older transistor horizontal-output arrangement is also covered. Pincushion circuitry is discussed. In addition, the obsolescent (but often-encountered) tube-type horizontal-sweep circuitry is included, with notes on common trouble symptoms and their causes.

The eighth chapter explains troubleshooting with the oscilloscope in the vertical-sweep section. Principles of vertical-sweep circuit action are presented, with examples of malfunction due to defective components or devices. Pincushion correction is described and illustrated. The theory behind generation of 50-volt vertical-circuit waveforms with a supply voltage of 11 volts is included. The chapter concludes with a practical review of troubleshooting tube-type vertical-sweep circuitry. In Chapter 9, signal tracing in the sound-if and audio section is detailed. Principles of intercarrier-sound signal processing are provided, and oscilloscope application techniques are explained. Causes of ratio-detector frequency distortion are included, and control of sync-buzz interference is discussed. Troubleshooting with the oscilloscope in the audio section is described, with particular attention being paid to distortion analysis. The chapter concludes with a review of troubleshooting tube-type intercarrier-sound systems.

In the final chapter the *digital revolution* has been given appropriate recognition. All troubleshooting with the oscilloscope was based on waveform analysis and interpretation prior to the popularization of digital computers and digital equipment in general. Today, however, the general electronics troubleshooter is likely to be concerned with both time/frequency and data-domain oscilloscope displays. In Chapter 10 I have tried to provide an adequate introduction to these means of troubleshooting digital electronics.

Electronics technology is becoming increasingly advanced. We technicians must keep up with these advances if we are to remain competitive. Unless the full capabilities of oscilloscope use are clearly understood, it will become difficult in the future for you to service modern circuitry properly. In preparing this new edition of *Troubleshooting With the Oscilloscope,* I have recognized this need and have made a dedicated effort to meet it. I strongly suggest that you work with your equipment as the various procedures are described. This "reinforced learning," gained at the workbench, will be much more valuable to you than the knowledge you can acquire from just reading the book.

ROBERT G. MIDDLETON

Contents

CHAPTER 1

INTRODUCTION TO THE OSCILLOSCOPE 11
Two Areas of Oscilloscope Application — Time/Frequency vs.
Data-Domain Oscilloscopes — Basic Oscilloscope Waveforms —
Oscilloscope Test Classifications — "Intelligent" Oscilloscopes

CHAPTER 2

HOW TO OPERATE AN OSCILLOSCOPE 41
Basic Oscilloscope With Free-Running Sweep — Survey of Control
Functions and Terminal Facilities — Vertical and Horizontal Lin-
earity — Peak-to-Peak and Peak Voltage Measurements — DC
Voltage Components and Measurements — AC Waveforms With
DC Components — Sync Function — Action of Triggered-Sweep
Controls — Dual-Trace Display Modes

CHAPTER 3

USING OSCILLOSCOPE PROBES 81
Why Probes Are Needed in Oscilloscope Tests and Measurements
— Low-Capacitance-Probe Construction and Adjustment — Construction and Operation of Demodulator Probes — Resistive "Isolating" Probe — High-Voltage Capacitance-Divider Probe — Overview
of Probe Application in TV Circuitry — Vertical Interval Test Signal — Special Types of Oscilloscope Probes

CHAPTER 4

SIGNAL TRACING IN RF, IF, AND VIDEO AMPLIFIERS . . . 105
Troubleshooting RF Amplifiers — Signal Tracing in the IF Section
— Signal Tracing in the Video Amplifier — Television Station Interference — Integrated-Circuit IF Systems — Transistor Replacement

CHAPTER 5

SIGNAL TRACING IN THE SYNC SECTION 141
  Oscilloscope Application Techniques and Pattern Evaluation —
  Color-Sync Troubleshooting With the Oscilloscope — Sync Trouble-
  shooting in Older Receivers — Note on Modular Construction

CHAPTER 6

TROUBLESHOOTING THE AFC AND HORIZONTAL-OSCILLATOR
  SECTION 165
  Overview of Horizontal-AFC Circuit Action and Waveforms —
  Pattern-Generator Sync Waveforms — Troubleshooting Tube-Type
  AFC and Horizontal-Oscillator Circuits

CHAPTER 7

WAVEFORM TESTS IN THE HORIZONTAL-SWEEP SECTION . . 181
  Troubleshooting the SCR Horizontal-Sweep System — Transistor
  Horizontal-Output Arrangement — Troubleshooting Tube-Type
  Horizontal-Sweep Circuitry

CHAPTER 8

TROUBLESHOOTING THE VERTICAL-SWEEP SECTION 199
  Principles of Vertical-Sweep Operation — Troubleshooting Tube-
  Type Vertical-Sweep Circuitry

CHAPTER 9

SIGNAL TRACING IN THE SOUND-IF AND AUDIO SECTIONS . . 217
  Principles of Intercarrier-Sound Signal Processing — Troubleshooting Tube-Type Intercarrier-Sound Systems

CHAPTER 10

DIGITAL-LOGIC TROUBLESHOOTING WITH THE OSCILLOSCOPE . 229
  Digital-Logic Waveform Relationships — Operating Waveforms in
  Simple Digital Networks — Data-Domain Displays

INDEX 249

CHAPTER 1

Introduction to the Oscilloscope

Prior to the digital revolution all servicing with the oscilloscope was based on waveform analysis and interpretation. Today, data-domain analysis and interpretation is also used.

TWO AREAS OF OSCILLOSCOPE APPLICATION

Analog circuitry (which dominates tv receiver operation), for example, is associated with waveforms such as shown in Fig. 1-1A. On the other hand, digital circuitry, which is included in many consumer electronics products and which is the "heart" of personal computers, is characterized by data displays, as exemplified in Fig. 1-1B. Events in a digital data stream are not referenced to real time; instead, they are referenced to clock time. (The "clock" is a high-frequency square-wave generator that synchronizes the functions in a digital system.)

A dual-trace oscilloscope is illustrated in Fig. 1-2. It provides time/frequency displays. This type of oscilloscope also provides vectorgram displays. A vectorgram is not a time/frequency pattern in a strict technical sense; instead, it is a voltage-voltage, voltage-current, or current-current display. These distinctions will be discussed subsequently.

TIME/FREQUENCY OSCILLOSCOPES VS. DATA-DOMAIN OSCILLOSCOPES

A high-performance data-domain type of oscilloscope is illustrated in Fig. 1-3. This particular instrument provides a choice of data-domain or time/frequency displays. Each type of display has its place in digital servicing procedures. The essential fact to be noted is that time/frequency displays alone are no longer adequate for troubleshooting digital circuitry—they must be supplemented by data-domain displays, particularly in preliminary procedures. The reason for this requirement is shown in Fig. 1-4; digital signal flow is ordinarily multiline, and the data stream must often be checked line by line to detect the malfunction and to close in on the defective device, faulty connection, or other "bug." After the problem area has been revealed, pinpointing the defect is often facilitated by changing over to time/frequency displays. Observe that display of a pattern such as shown in Fig. 1-1B requires an *oscilloscope with a RAM* (*read-and-write memory*). A selected portion of the data stream is entered into the RAM. Integrated-circuit (IC) character generators are then driven by this memory to display the corresponding data domain on the oscilloscope screen.

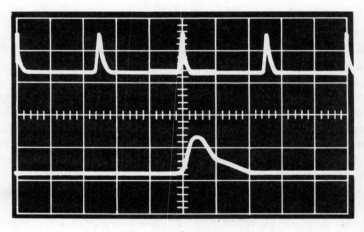

(A) Dual-trace time/frequency pattern.

(C) Typical integrated circuit used in character generation.

Fig. 1-1. Examples of time/frequency

BASIC OSCILLOSCOPE WAVEFORMS

A sine wave, shown in Fig. 1-5, is the most basic waveform. We describe a sine wave in terms of degrees and volts or amperes. The *peak voltage* of a sine wave is measured from the 0-volt (reference) line to the point corresponding to 90° on the positive half-cycle or to the point corresponding to 270° on the negative half-cycle. The *peak-to-peak voltage* is measured from the positive peak at 90° to the negative peak at 270°. The *rms voltage* of a sine wave is equal to 0.707 of its peak voltage. The *average* value of one-half cycle of a sine wave is equal to 0.318 of the peak value; the average value of two half-cycles is equal to 0.637 of peak. As shown in Fig. 1-6 the average

1601L LOGIC
STATE ANALYZER
HEWLETT PACKARD

BITS BCD	11 10 9	8 7 6	5 4 3	2 1 0	
	1 0 0	0 0 0	1 0 0	0 0 0	1
	1 0 0	0 0 1	1 0 0	0 1 0	2
	0 0 0	0 1 0	0 0 1	1 0 0	3
	0 0 0	0 1 1	0 0 1	1 1 0	4
	0 0 0	0 1 1	0 0 1	1 1 1	5
	1 0 0	0 0 0	1 0 0	0 0 0	6
	1 0 0	0 0 0	1 0 0	0 0 1	7
	1 0 0	0 0 1	1 0 0	0 1 0	8
	0 0 0	0 1 0	0 0 1	1 0 0	9
	0 0 0	0 1 1	0 0 1	1 1 0	10
	0 0 0	0 1 1	0 0 1	1 1 1	11
	1 0 0	0 0 0	1 0 0	0 0 0	12
	1 0 0	0 0 0	1 0 0	0 0 1	13
	1 0 0	0 0 1	1 0 0	0 1 0	14
	0 0 0	0 1 0	0 0 1	1 0 0	15
	0 0 0	0 1 1	0 0 1	1 1 0	16
BITS OCT	11 10 9	8 7 6	5 4 3	2 1 0	

(B) Digital-word pattern, displayed with reference to clock time.

and data-domain oscilloscope displays.

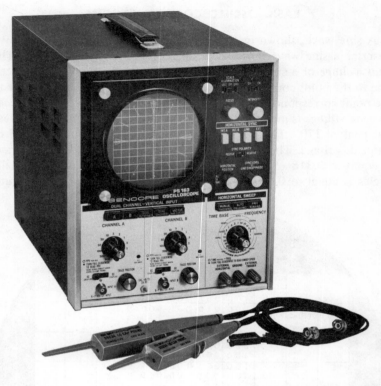

Fig. 1-2. A widely used type of dual-trace oscilloscope.

value of a half-sine wave is of importance in half-wave rectification. Thus when an oscilloscope is operated on its dc input function the peak voltage of the (half-wave) waveform can be measured by counting the squares on the graticule from the 0-volt axis to the peak of the waveform. On the other hand, if an oscilloscope is operated on its ac input function, the half-rectified waveform will be displayed with the 0-volt axis passing through the average value of the waveform (dotted line in Fig. 1-5). This fact can be stated as a law:

> Whenever any waveform is displayed on the screen of an oscilloscope with ac input coupling (*RC*-coupling), the pattern will position itself vertically on the screen with its positive-peak voltage above the 0-volt axis, and with its negative-peak voltage below the 0-volt axis.

As will be explained in greater detail the vertical amplifier in an oscilloscope is ordinarily calibrated in peak-to-peak volts per centimeter (V/cm) of vertical deflection. It follows from the relations shown in Fig. 1-5 that peak voltage is measured in the same calibration units as peak-to-peak voltage. Therefore, if an oscilloscope has been calibrated to measure peak-to-peak volts, this calibration is also valid for measuring peak volts. Note in Fig. 1-6 that when

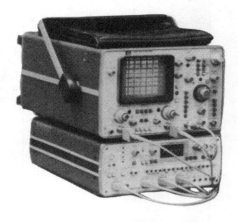

Courtesy Hewlett-Packard, Inc.

Fig. 1-3. A high-performance digital-logic analyzer.

ac input coupling is used, area A_1 will be exactly equal to area B_1. This equality results from the fact that there is just as much electrical quantity in the positive-peak excursion as there is in the negative-peak excursion of a waveform. Thus the average value of a full sine-wave cycle is zero.

We can also describe a sine wave in terms of time. Thus if one complete waveform or 360° are completed in one microsecond, the wave is said to have a frequency of one million hertz, or one megahertz (1 MHz). The term "hertz" (abbreviated Hz) means *cycles per second*. We will find that sine waves, however, are in the minority in the vast array of waveforms encountered at the service bench. For example, pulse waveforms, such as sync pulses, keying pulses, and blanking pulses, occur in tv circuitry. The basic features of an ideal pulse waveform are shown in Fig. 1-7. If the waveform is displayed with ac-coupled input to the oscilloscope, its positive excursion will appear above the zero-reference line on the oscilloscope screen, and its negative excursion will appear below the zero-reference line. The *pulse width* is measured in time units, such as microseconds. The *pulse repetition time* (period) is also measured in time units (with the aid of an oscilloscope that has triggered sweeps and a calibrated time base). The reciprocal of the pulse repetition time

DATA STREAM APPLIED TO DIGITAL LOGIC ANALYZER

Start Display, Trigger Mode, No Delay. In this Mode the Trigger Word and the Next 15 Words are Displayed.	End Display, Trigger Mode, No Delay. The 15 Words Preceding the Trigger Word are Displayed.	End Display with 1 to 15 Clocks of Digital Delay. By Selecting End Display and Digital Delay up to 15, it is Possible to Simultaneously Display Words Ocurring Before and After the Trigger Word.	Start Display with Digital Delay. The Trigger Word is Not Displayed in this Mode. After the Preselected Delay a Display Field of 16 Words is Displayed.

Courtesy Hewlett-Packard, Inc.

Fig. 1-4. Data stream analysis to locate trouble is greatly facilitated by data-domain oscilloscope displays.

is equal to the pulse repetition rate (sometimes called the *pulse frequency*). The *duty cycle* is equal to the pulse width divided by the pulse repetition time. A constant voltage may be regarded as a limiting case of the pulse waveform for which the duty cycle is equal to unity.

Although pulses and square waves have no apparent similarity to sine waves, it should be noted that the sine wave remains the basic

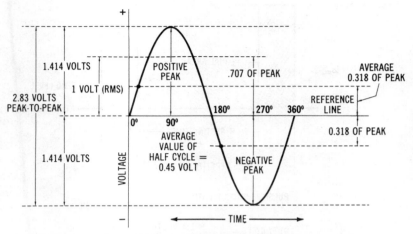

Fig. 1-5. Voltage relations in a pure sine wave.

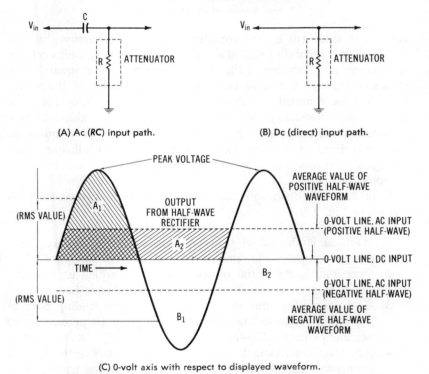

(A) Ac (RC) input path.

(B) Dc (direct) input path.

(C) 0-volt axis with respect to displayed waveform.

Fig. 1-6. Oscilloscope ac and dc input coupling and resulting waveform display.

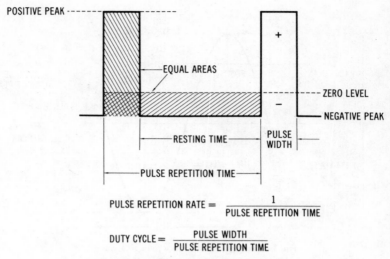

$$\text{PULSE REPETITION RATE} = \frac{1}{\text{PULSE REPETITION TIME}}$$

$$\text{DUTY CYCLE} = \frac{\text{PULSE WIDTH}}{\text{PULSE REPETITION TIME}}$$

Fig. 1-7. Basic features of an ideal pulse waveform.

element in these (as in all) complex waveforms. In many practical situations it is helpful to regard a square wave as being built up from sine waves, as illustrated in Fig. 1-8. For example, if a square wave is passed through a narrow-band circuit, the corners of the waveform become rounded, which we may regard as resulting from loss of the higher-frequency sine-wave components. Note also that this distortion is accompanied by a slowed *rise* and *fall* of a square or pulse waveform. Again, this increased rise time (and fall time) may be regarded as resulting from a loss of the higher-frequency sine-wave components.

Note the basic distortion factors in a digital clock waveform, as shown in Fig. 1-9:

1. *Preshoot* is the initial transient response to a unidirectional change in input, which precedes the main excursion.
2. *Overshoot* is the initial transient response, which exceeds the steady-state response, to a unidirectional change in input.
3. *Rise time* is the time that is required for the leading edge of the waveform to increase in amplitude from 10 percent to 90 percent of its final value.
4. *Rounding* denotes lack of a sharp corner in a waveform, or of a smooth transition from the leading edge to the trailing edge to the limiting final value.

5. *Ringing* denotes a damped oscillation in the output signal in response to a sudden change in the input signal.
6. *Settling time* is equal to the time interval, following a specified excitation of a system, which is required for a particular variable to enter and to remain within a designated narrow band centered on the final value of the variable.
7. *Sag* denotes the decrease in mean waveform amplitude, expressed as a percentage of its full amplitude, at a specified time following the initial attainment of full amplitude.
8. *Fall time* is the time that is required for the trailing edge of the waveform to decrease in amplitude from 90 percent to 10 percent of its final value.
9. *Base line offset* denotes the difference between the desired value or condition, and the value or condition that is actually obtained.

Any of these distortion factors may occur singly or in combination, as exemplified in Fig. 1-11. Any single distortion or multiple distortion may be slight or may be substantial. All waveshapes have a design *tolerance;* slight distortions are evaluated as being within permissible tolerance, whereas substantial distortions are considered to be symptoms of trouble.

"Hidden" Distortion

Distortion factors that are *completely invisible on slow-speed sweep can become apparent on high-speed sweep.* Consider, for example, the 20-μs (microsecond) pulse depicted in Fig. 1-11A. This pattern indicates that the waveform has sharp corners and zero rise time. An expansion of the display of the pulse at higher sweep speeds would be made as shown in Figs. 1-11B, 1-11C, 1-11D, and 1-11E. In this time-scale expansion the waveform still has sharp corners and zero rise time at the highest sweep speed (0.04 μs/cm; see Fig. 1-11E). Next, let us see how this 20-μs pulse is actually displayed in practice at higher sweep speeds. With reference to Fig. 1-12 we observe that the pulse does not really have perfectly sharp corners, and that its rise time is greater than zero. Thus, at a sweep speed of 0.04 μs/cm (Fig. 1-12E) it is apparent that the corners of the pattern are actually rounded and that the rise time of the leading edge is approximately 0.04 μs. Thus, the *limited resolution* of the human eye causes smaller distortion factors to be invisible on slow-speed sweep, whereas these distortion factors emerge into clear visibility on a high-speed sweep.

Rise time, per se, is not a form of distortion, although *excessive* rise time is regarded as distortion factor; excessive rise time points to an associated circuit malfunction. As shown in Fig. 1-13 the

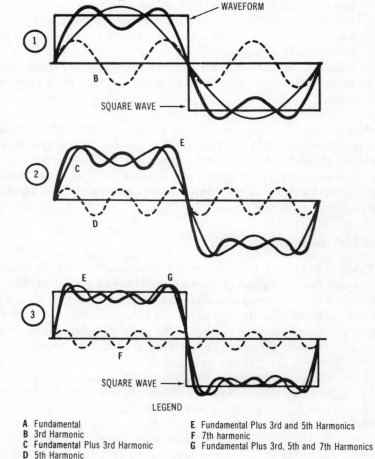

LEGEND

A Fundamental
B 3rd Harmonic
C Fundamental Plus 3rd Harmonic
D 5th Harmonic

E Fundamental Plus 3rd and 5th Harmonics
F 7th harmonic
G Fundamental Plus 3rd, 5th and 7th Harmonics

(A) Components of a square wave.

Fig. 1-8. Build-up of a

rise time is measured from the 10-percent point to the 90-percent point on the leading edge of a pulse waveform. As noted previously, the rise time may not be measurable, and its presence may be invisible until the pulse pattern has been greatly expanded on a triggered-sweep time base. Apprentice technicians are sometimes puzzled by the fact that the leading edge of a square wave or pulse may be completely invisible when the waveform is displayed on slow-speed sweep. This is a result of the comparatively swift travel of the electron beam during the rise interval, compared with its travel speed along the flat top of the waveform. If the sweep speed is increased, however, the leading edge becomes visible because the scope operator advances the intensity control. The intensity control can be advanced in this situation without burning the crt screen because the leading edge of the waveform becomes inclined (slanted) at high sweep speed, and the electron beam travels at a more nearly uniform rate along the leading edge and along the flat top of the pattern.

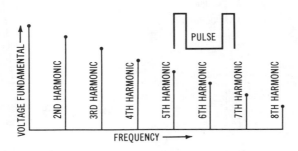

(B) Harmonic components in a pulse waveform.

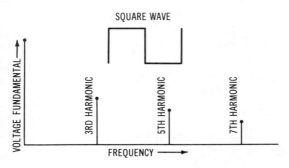

(C) Harmonic components in a square wave.

square wave from sine waves.

Natural Law of Growth and Decay

The exponential waveform (Fig. 1-14) exhibits the natural law of growth and decay, and it is one of the fundamental waveforms. When a square wave or pulse waveform is passed through an *RC* differentiating circuit, the output waveform has exponential leading and trailing edges. Observe in Fig. 1-14 that the time constant of the differentiating circuit can be measured with a triggered-sweep scope that has a calibrated time base. The time constant of an *RC* circuit is measured as the product of resistance and capacitance. For example,

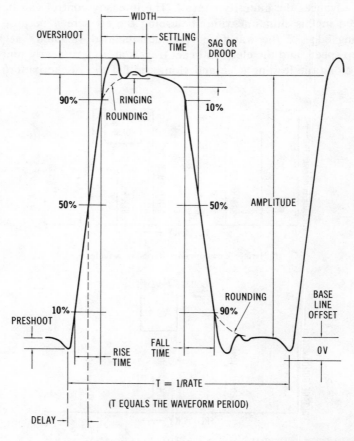

Courtesy Hewlett-Packard, Inc.

Fig. 1-9. Basic distortion factors in a digital clock waveform.

if a 1-megohm resistor is connected to a 1-microfarad capacitor, the time constant of this *RC* circuit is 1 second. With reference to Fig. 1-14A it will then take 1 second for the leading edge of the output waveform to attain 63.2 percent of its final value. Similarly, it will

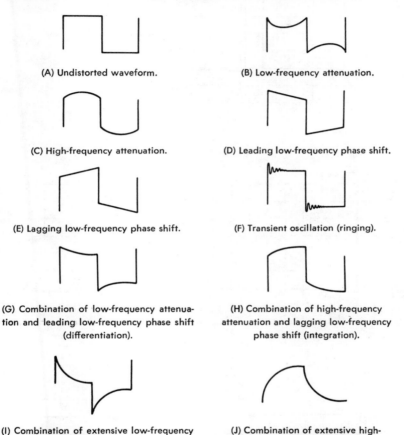

(A) Undistorted waveform.

(B) Low-frequency attenuation.

(C) High-frequency attenuation.

(D) Leading low-frequency phase shift.

(E) Lagging low-frequency phase shift.

(F) Transient oscillation (ringing).

(G) Combination of low-frequency attenuation and leading low-frequency phase shift (differentiation).

(H) Combination of high-frequency attenuation and lagging low-frequency phase shift (integration).

(I) Combination of extensive low-frequency attenuation and leading low-frequency phase shift (differentiation).

(J) Combination of extensive high-frequency attenuation and lagging low-frequency phase shift (integration).

Fig. 1-10. Distortion factors may occur singly or in combination.

then take 1 second for the trailing edge of the output waveform to fall to 36.8 percent of its initial amplitude. If the pulse response of a circuit with nonlinear resistance(s) is displayed, we may find that the rise time of the output waveform is different from its fall time, as shown in Fig. 1-15.

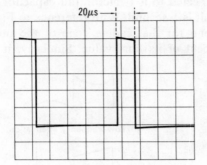

(A) Sweep at 0.02 ms/cm.

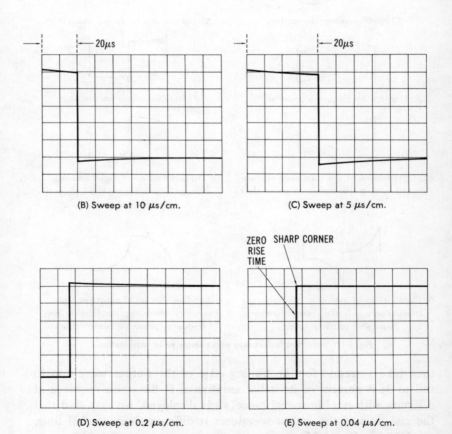

(B) Sweep at 10 μs/cm.

(C) Sweep at 5 μs/cm.

ZERO RISE TIME   SHARP CORNER

(D) Sweep at 0.2 μs/cm.

(E) Sweep at 0.04 μs/cm.

Fig. 1-11. Pulse waveform of 20-μs width progressively expanded on triggered sweep.

Of course, the troubleshooter generally encounters more complex circuitry than simple *RC* differentiators and integrators. For example, a two-section integrator or a three-section integrator may be "pack-

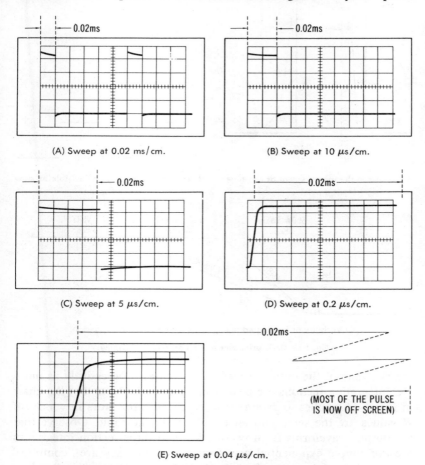

(A) Sweep at 0.02 ms/cm.

(B) Sweep at 10 μs/cm.

(C) Sweep at 5 μs/cm.

(D) Sweep at 0.2 μs/cm.

(E) Sweep at 0.04 μs/cm.

Fig. 1-12. Progressive expansion of a 20-μs pulse on the screen of a triggered-sweep scope.

aged" in an encapsulated unit. The oscilloscope operator must therefore check a pulse or square waveform after it has passed through two or three *RC* sections connected in series. Fig. 1-16 shows a universal time-constant chart for leading edges of output waveforms from one-, two-, and three-section integrators. (Trailing edges are the same as the leading edges "turned upside down.") As would be expected,

25

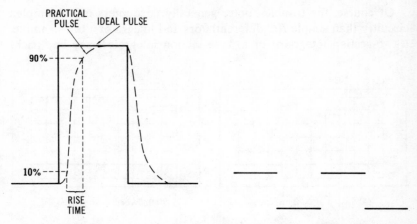

(A) Factors in the measurement of rise time of a pulse.

(B) Leading edge is often invisible at slow sweep speed.

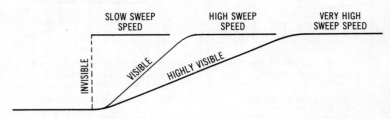

(C) Visibility of leading edge improves with waveform expansion.

Fig. 1-13. Basic principles of triggered-sweep display.

the rise time of the output waveform increases progressively as more *RC* integrating sections are connected in cascade. Note that the chart in Fig. 1-16 applies to *symmetrical* cascaded integrators in which all *R* values are the same and all *C* values are the same. Observe that the output waveforms from two-section and three-section integrators are *not* simple exponential waveforms. Instead, they are "complex" exponentials, due to the fact that a following *RC* section loads a preceding *RC* section—since an *RC* section is frequency selective (has filter action), it is not a resistive load.

Steady-State vs. Transient Response

A sine waveform, as from a signal generator, represents a steady-state test signal. Technically, a steady-state waveform has only one frequency. By way of comparison a square-wave signal represents a

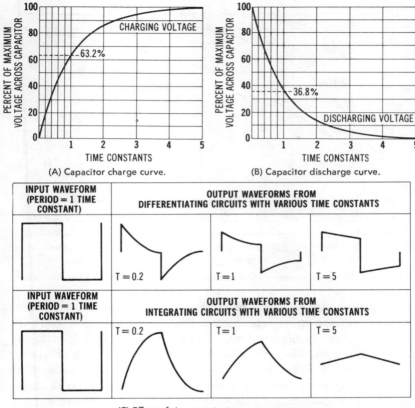

(A) Capacitor charge curve.

(B) Capacitor discharge curve.

(C) Effect of time constant on square wave.

Fig. 1-14. Universal _RC_ time-constant charts.

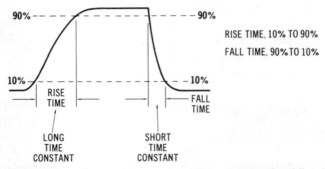

RISE TIME, 10% TO 90%

FALL TIME, 90% TO 10%

Fig. 1-15. Circuits with nonlinear resistance may exhibit a rise time that differs from the fall time.

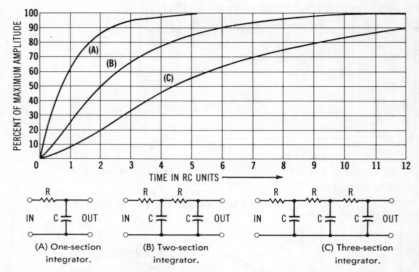

Fig. 1-16. Universal time-constant chart for one-, two-, and three-section integrator circuits.

transient test signal; it has an extensive array of harmonic frequencies, in addition to its fundamental frequency (repetition rate). In practice, signal waveforms are used which have both steady-state and transient aspects. For example, burst waveforms consist of groups of sine waves. Within a group the waveform represents a steady-state condition; however, burst waveforms are associated with an extensive array of harmonic frequencies, inasmuch as each successive burst is keyed (switched) on and off. In this respect a burst waveform is regarded as a transient test signal. Basic types of burst waveforms are shown in Fig. 1-17. Burst frequencies range from the lower audio

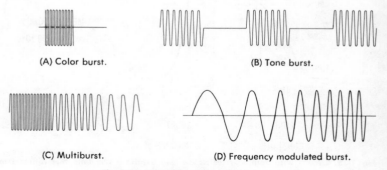

(A) Color burst.

(B) Tone burst.

(C) Multiburst.

(D) Frequency modulated burst.

Fig. 1-17. Typical burst waveforms.

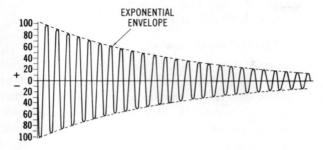

Fig. 1-18. Typical ringing waveform, as displayed on oscilloscope screen.

region to the intercarrier sound frequency (4.5 MHz) in troubleshooting with the oscilloscope.

Modified Sine Waveforms

Another basic type of transient waveform is the damped sine wave, exemplified in Fig. 1-18. It is also called a *ringing waveform,* and it will be encountered in many types of circuitry. Note that a damped sine wave does not have a single frequency—it is associated with harmonics, not only because it starts abruptly, but also because its amplitude decreases exponentially. The ringing frequency of an *RLC* circuit or system is termed its *natural frequency.* The rate at which the waveform decays is a measure of the quality or *Q* of the circuit or system. Ringing waveforms are produced by shock excitation (impulse excitation) of an *RLC* circuit or system, as by square-wave or pulse testing. Applications for ringing-waveform tests will be ex-

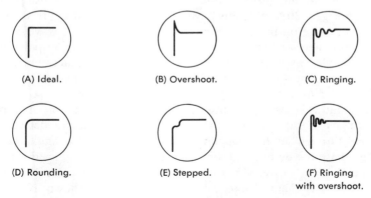

(A) Ideal.   (B) Overshoot.   (C) Ringing.

(D) Rounding.   (E) Stepped.   (F) Ringing with overshoot.

Fig. 1-19. Basic modes of cornering.

plained subsequently. One such application with respect to system frequency response is described next.

Cornering of Pulses and Square Waves

Rise time and fall time are measured between the 10-percent and the 90-percent points on the leading and trailing edges of a waveform because this convention minimizes or eliminates errors from cornering deviations. Referring to Fig. 1-19, there are five basic forms of corner distortion. *Overshoot* is often caused by uncompensated capacitance in the circuit under test; it can be regarded as residual system differentiation of a square wave or pulse waveform. *Ringing* is caused by lack of uniform frequency response; it usually indicates a rising high-frequency response, followed by a falling high-frequency characteristic. Ringing may also be the result of a "notch" in a frequency response curve. Corner *rounding* is associated with inadequate high-frequency response with respect to the test waveform (pulse or square wave). A *stepped* corner response is commonly encountered in multistage compensated video amplifiers (it can occur in oscilloscope vertical amplifiers). If one stage has falling high-frequency response and is followed by a stage with rising high-frequency response, a pulse may be reproduced with stepped corner distortion. *Ringing with overshoot* is also commonly encountered in multistage compensated video amplifiers. If one stage develops overshoot and is followed by a stage that tends to ring, the result is overshoot with ringing in a reproduced pulse or square waveform.

Pulse Polarities

Troubleshooting procedures are often concerned with pulse polarities. In Fig. 1-20A, a positive sync pulse is displayed entirely above the 0-volt line on the oscilloscope screen. This type of waveform is obtained when a picture-detector diode is correspondingly polarized, for example. Suppose, on the other hand, that a replacement diode is connected into the circuit with reverse polarity; then the display will appear as shown in Fig. 1-20B. A negative sync pulse is produced, and it is displayed entirely below the 0-volt line on the scope screen. Next, suppose that the output from the picture detector is passed through an *RC* coupling circuit into the video amplifier. In such a case the dc pulse output from the detector is changed into an ac pulse input to the video amplifier. Thus the dc pulse shown in Fig. 1-20A is changed into the ac pulse waveform shown in Fig. 1-20C. This is called a *positive-going sync pulse of the ac type.* Simi-

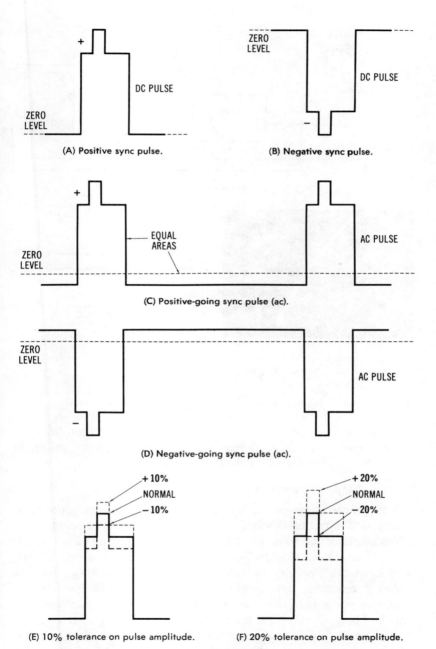

(A) Positive sync pulse.

(B) Negative sync pulse.

(C) Positive-going sync pulse (ac).

(D) Negative-going sync pulse (ac).

(E) 10% tolerance on pulse amplitude.

(F) 20% tolerance on pulse amplitude.

Fig. 1-20. Ac and dc pulses.

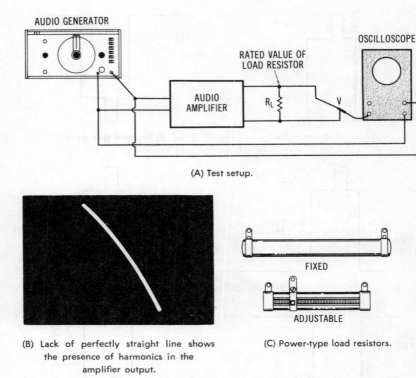

(A) Test setup.

(B) Lack of perfectly straight line shows the presence of harmonics in the amplifier output.

(C) Power-type load resistors.

Fig. 1-21. A basic audio-frequency Lissajous pattern development.

larly, if the dc pulse in Fig. 1-20B is passed through an *RC* coupling circuit, it is changed into the ac pulse waveform shown in Fig. 1-20D, and is called a *negative-going ac sync pulse*.

Lissajous Family Waveforms

Various categories of waveforms in the Lissajous family are commonly used in troubleshooting with the oscilloscope. For example, an audio amplifier can be checked for distortion as shown in Fig. 1-21A. Lack of a perfect straight-line pattern shows the presence of harmonics in the amplifier output. Many types of distortion, such as clipping, crossover, and nonlinear distortion, can be identified in Lissajous patterns by an experienced troubleshooter. Another category of Lissajous pattern is exemplified in Fig. 1-22. This is a vectorscope test, in which a keyed rainbow generator signal is applied to a color-tv receiver, and the signals from the red and blue guns of

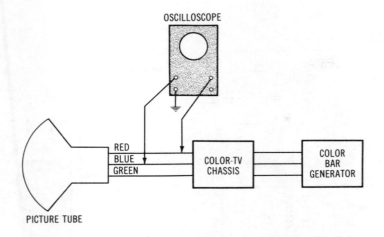

(A) Test setup.

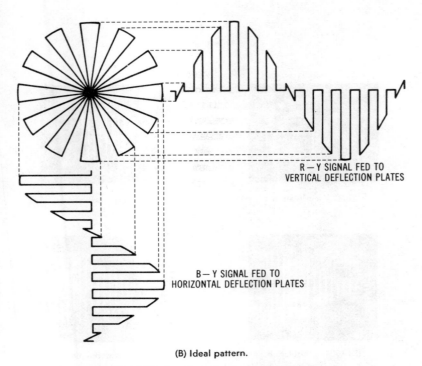

R — Y SIGNAL FED TO
VERTICAL DEFLECTION PLATES

B — Y SIGNAL FED TO
HORIZONTAL DEFLECTION PLATES

(B) Ideal pattern.

Fig. 1-22. A Lissajous figure (vectorgram) development.

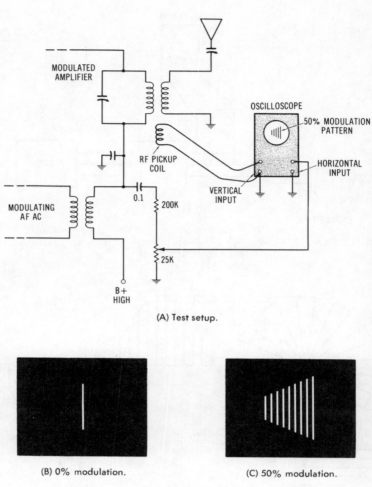

(A) Test setup.

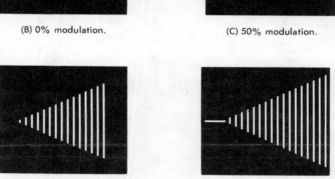

(B) 0% modulation.

(C) 50% modulation.

(D) 100% modulation.

(E) Overmodulation.

Fig. 1-23. Lissajous patterns show percentage of modulation.

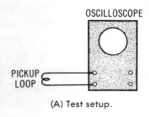

(A) Test setup.

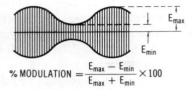

$$\% \text{ MODULATION} = \frac{E_{max} - E_{min}}{E_{max} + E_{min}} \times 100$$

(B) Waveform observed when transmitter is energized by a tone signal.

Fig. 1-24. Percentage modulation of an am waveform.

the picture tube are fed to the vertical and horizontal inputs of an oscilloscope. This type of test is particularly useful in checking for correct subcarrier injection phases.

Lissajous figures are also used to check for percentage of amplitude modulation, as in CB transceivers. The test setup is shown in Fig. 1-23A. An rf pickup coil is used to apply the modulated rf signal to the vertical input of the oscilloscope, and a potentiometer is connected at the output of the modulator to feed audio modulating signal into the horizontal input of the oscilloscope. At 50-percent modulation, a trapezoidal pattern is displayed on the oscilloscope screen. At zero modulation a straight vertical line is displayed. At 100-percent modulation a perfect triangle is displayed (unless there is distortion occurring in the modulation process). When overmodulation is present, a "tail" is appended to the triangular pattern. Compare this test method with that shown in Fig. 1-24, in which the modulated-rf waveform is displayed on sawtooth sweep. The same E_{max} and E_{min} equation applies to trapezoidal modulation patterns also. In other words, the left-hand vertical deflection represents E_{min}, and the right-hand vertical deflection represents E_{max} in the trapezoidal patterns.

OSCILLOSCOPE TEST CLASSIFICATIONS

Any troubleshooting procedure with the oscilloscope can be classified under one or more of the following basic tests.

1. *Signal tracing.* This procedure consists in checking for the presence or absence of a signal voltage at successive points through a signal channel. Its capability is limited only by the frequency response and the sensitivity of the vertical amplifier of the oscilloscope. Thus an audio-frequency, radio-frequency, intermediate frequency, or video-frequency signal may be checked.

2. *Waveform checks.* Waveform checks include amplitude (peak-to-peak voltage) measurement and waveshape observations. These data, in turn, are compared with specifications in the receiver service data. Waveform checks are quantitative, whereas signal-tracing tests are qualitative. Quantitative measurements are subject to reasonable tolerances. For example, a permissible amplitude variation of ±10, or ±20 percent may be noted in the service data.

3. *Frequency response.* Sweep and marker generators are used with the oscilloscope for visual checks of frequency response. The displayed patterns are termed *frequency-response curves.* A response curve shows the relative output voltages at all frequencies within the passband of the amplifier or circuit under test. This displayed response curve, in turn, is compared with specifications in the receiver service data.

4. *Phase checks.* A phase test displays a pattern (or patterns) on the oscilloscope screen showing the phase difference between a reference signal and the signal being checked. Specialized test signals may be employed, particularly in color-tv troubleshooting procedures. Suitable test signals are specified in receiver service data.

5. *Frequency measurements.* Frequency measurements are made to good advantage with a triggered-sweep scope that has a calibrated time base. For example, the accuracy of a square-wave generator or an audio oscillator can be closely checked in this manner. This method measures the fundamental frequency of a waveform. When a complex waveform is being checked, it is advantageous to use a specialized type of oscilloscope, called a *spectrum analyzer* (Fig. 1-25). For example, this test shows the frequency and amplitude of each distortion product in the output waveform from an audio amplifier.

6. *Time measurement.* Time measurement is ordinarily made with a triggered-sweep oscilloscope that has a calibrated time base. This type of test is used to check pulse widths in solid-state horizontal-sweep circuits, in blanking circuits, and in tv digital circuitry.

7. *Transient tests.* Pulse, square-wave, and tone-burst generators are used with the oscilloscope for making transient tests. A multiburst generator is used with a supplementary oscilloscope for combination transient and frequency response checks of overall tv receiver response.

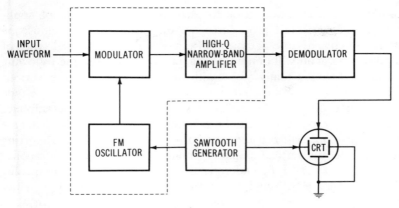

(A) Analyzer circuit.

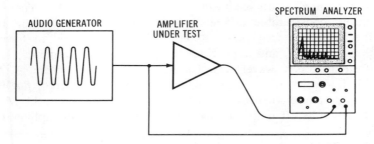

(B) Test setup.

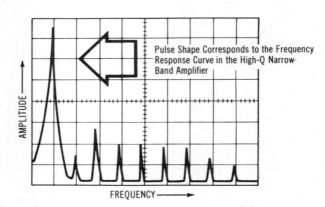

(C) Display of amplifier waveform frequency response.

Fig. 1-25. A spectrum analyzer shows the frequency and amplitude of each harmonic or other component of a complex waveform.

With this brief introduction to oscilloscopes, their areas of application, and basic test procedures, we turn next to the oscilloscope itself and consider the operation of controls in the principal types of oscilloscope. You will obtain the maximum value from the following chapters if you actually work with an oscilloscope as the various techniques are described. Note in passing that modern service-type oscilloscopes are not standardized with respect to vertical-amplifier frequency response. For troubleshooting color-tv circuitry, only a wideband scope will provide undistorted chroma patterns. Referring to Fig. 1-26, the vertical amplifier should provide frequency-response curves such as *A* or *B*. Older types of oscilloscopes had frequency responses such as *C* and *D;* they are unsuitable for color-tv troubleshooting.

Observe in Fig. 1-26 that the *A* and *B* frequency-response curves have the same −3-dB bandwidth: approximately 11 MHz. There is, however, considerable difference in the roll-off rates of the two vertical-amplifier frequency responses. Curve *A* is said to have a *Gaussian* roll-off, and it provides the best transient response (least distortion) for tv waveforms that have higher harmonics past the cutoff frequency of the oscilloscope. Curve *B* is called a *flat* frequency response—it provides somewhat more gain up to 11 MHz, and then rolls off quite rapidly. This type of vertical-amplifier fre-

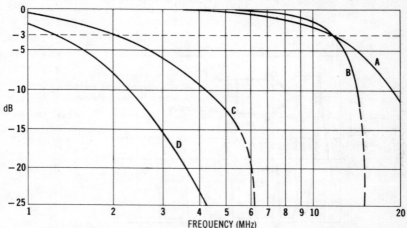

A -Gaussian or Normal, Response—Gradual Rolloff (9dB/Octave). Rise Time Approximately 0.035 Microsecond.
B -Flat Response to 5 or 6 MHz. More Rapid Rolloff Than Curve A. Rise Time Approximately 0.035 Microsecond.
C -Old IRE Curve. Rise Time 0.175 Microsecond.
D -New IRE Curve. Rise Time 0.3 Microsecond.

Fig. 1-26. Oscilloscope response curves.

quency response provides least distortion of the color-burst signal. Curve *C* is representative of many older designs of black-and-white service oscilloscopes; it has a −3-dB bandwidth of 2 MHz, which is adequate for reproduction of the horizontal-sync pulse. Curve *D* is representative of obsolescent designs of black-and-white service scopes; it has a −3-dB bandwidth of approximately 1.15 MHz, which is marginal for reproduction of the horizontal-sync pulse waveform.

"INTELLIGENT" OSCILLOSCOPES

Modern oscilloscopes of sophisticated design can be characterized as "intelligent" instruments, compared with older types of oscilloscopes. For example, the automatic sync feature of many modern oscilloscopes would have been regarded as an "impossible" mode of operation in former years. The automatic sync function serves to lock any pattern on the screen, without adjustment of any sync controls. Lab-type oscilloscopes may also be provided with crt readout; this function employs computer circuitry to spell out waveform information automatically on the oscilloscope screen. As an illustration, crt readout will display alphanumeric characters on the screen that tell the oscilloscope operator what vertical sensitivity is being used, what horizontal-sweep speed is being used, whether the horizontal magnifier is in operation, whether a direct probe or a 10× probe is being used, which vertical or horizontal controls may be set to uncalibrated modes, and so on. The crt readout may be designed for display of pulse-width values, rise-time values, and repetition rates. Consequently, "intelligent" oscilloscopes with crt readout can "do the figuring" for the operator, reduce the time that is required for measurement, and avoid operator errors by displaying the basic data that are involved. Alphanumeric characters to be displayed on the crt screen are produced by integrated-circuit character generators under microprocessor control.

A *storage oscilloscope* can retain a pattern on its screen indefinitely. Even a brief nonrepetitive transient can be captured and "frozen" on the screen of a storage oscilloscope. This instrument is used chiefly in laboratories. It provides improved types of displays for observing changes in a signal, to illustrate the result of making a circuit adjustment, to compare the performance of two or more equipment setups, to compare an arbitrary waveform against a standard waveform, for reducing flicker in display of waveforms with low repetition rates, for effectively displaying digital signals that have a low duty cycle, for reducing the interfering effect of noise on a recurrent waveform, for unattended oscilloscopic monitoring of intermittent conditions, and for photography of multiple patterns in which time lapses are involved in obtaining individual patterns.

CHAPTER 2

How to Operate an Oscilloscope

It is sometimes supposed that oscilloscopes are difficult to operate because they have a comparatively large number of controls; a basic oscilloscope, such as depicted in Fig. 2-1, has about a dozen control knobs and switches. However, if the operation and response of each control or switch is considered step by step, an oscilloscope soon loses its mystery. Most service-type scopes are ac-operated and hence have a power cord which must be plugged into a 117-volt, 60-Hz outlet.

BASIC OSCILLOSCOPE WITH FREE-RUNNING SWEEPS

In the example of Fig. 2-1 the POWER ON-OFF switch is operated in conjunction with the intensity control. Accordingly, the INTENSITY control is turned from its "off" position to the right, just as a radio or television receiver is turned on. When power is applied to the os-cilloscope circuits, a pilot lamp glows or, in the case of an elaborate oscilloscope, an edge-lighted graticule is illuminated. When edge light-ing is used, the vertical and horizontal rulings glow, instead of appear-ing as black lines. (See Fig. 2-2.)

Intensity-Control Adjustment

After the oscilloscope has warmed up for a short time a glowing spot or line may appear on the screen. If it does not, turn up the INTENSITY control as required. CAUTION: Never advance the INTEN-

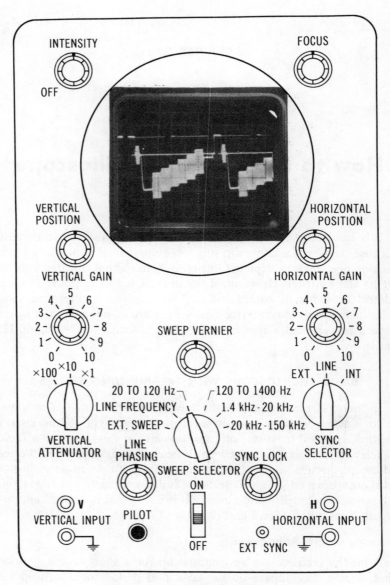

Fig. 2-1. Controls of a basic oscilloscope.

SITY control more than is required for a plainly visible spot or line on the screen—failure to observe this precaution can result in burning the crt screen. As in the case of a tv picture tube, screen damage is most likely to occur if the electron beam is forming a small, brilliant dot on the screen. Suppose that a spot or line does not appear on the scope screen, even when the INTENSITY control is fully advanced. In this case, check the settings of the HORIZONTAL and

Fig. 2-2. Graticule lines glow when edge lighting is provided.

VERTICAL POSITION (centering) controls. If either or both should be set considerably off center, the anticipated spot or line can be thrown off-screen.

Centering-Control Adjustment

With reference to Fig. 2-3, a spot display moves up or down the screen as the VERTICAL POSITION control is rotated back and forth. Or, the spot moves left or right on the screen as the HORIZONTAL POSITION control is rotated back and forth. Note in passing that *in theory any desired pattern could be (slowly) traced out on the screen by turning the positioning controls.* This is a simple manual analogy to the pattern development that takes place automatically when the electronic circuits of an oscilloscope are in operation. Referring to Fig. 2-4, the display of a geometric figure by a moving spot involves *persistence of vision.* This is the same phenomenon that is utilized to produce moving pictures from a series of still pictures. It is evident that the lower limit of the operator's persistence of vision corresponds to the lowest frequency (lowest sweep speed) that can be used without the pattern deteriorating into a meaningless moving spot.

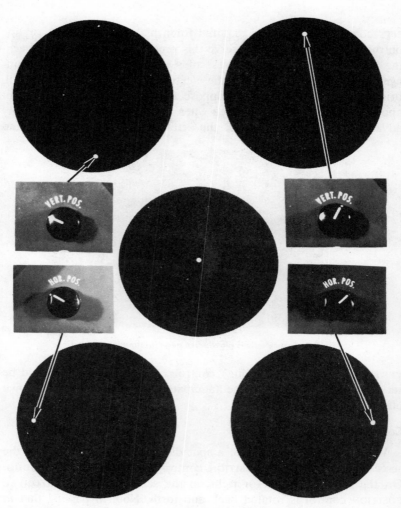

Fig. 2-3. Action of vertical and horizontal positioning (centering) controls.

Note, however, that *very low frequency* waveforms can be satisfactorily displayed if a crt screen phosphor is used that has unusually long persistence. In tv troubleshooting procedures a medium-persistence phosphor is preferred.

Focus-Control Adjustment

As in the case of a camera the crt in an oscilloscope must be properly focused to obtain a sharp pattern (the crt contains an electro-

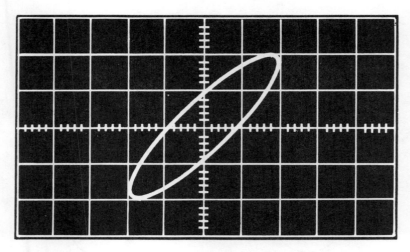

Fig. 2-4. Persistence of vision is involved when a moving spot "looks like" a geometric figure.

static lens). Fig. 2-5 shows how the appearance of a spot on the oscilloscope screen changes as the FOCUS control is turned. The FOCUS control is adjusted for the smallest spot possible. Note that in most oscilloscopes there is more or less interaction between the settings of the FOCUS and the INTENSITY controls. Therefore, it is quite possible that the FOCUS control may need to be readjusted if the INTENSITY-control setting is changed. Some elaborate oscilloscopes have an external (panel-mounted) ASTIGMATISM *control*. An ASTIGMATISM control is a focus "trimmer" that provides *uniform focus* at all points on the screen. A circular pattern can be displayed as shown in Fig. 2-6. When the FOCUS control is properly adjusted, the pattern will normally appear as in Fig. 2-6B. If the ASTIGMATISM control is out of

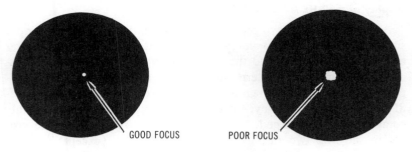

GOOD FOCUS   POOR FOCUS

Fig. 2-5. Action of FOCUS control.

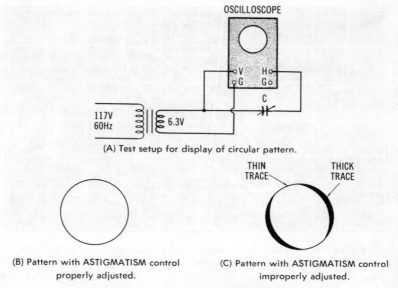

(A) Test setup for display of circular pattern.

(B) Pattern with ASTIGMATISM control properly adjusted.

(C) Pattern with ASTIGMATISM control improperly adjusted.

Fig. 2-6. FOCUS and ASTIGMATISM.

adjustment, however, the pattern will be displayed as shown in Fig. 2-6C.

SURVEY OF CONTROL FUNCTIONS AND TERMINAL FACILITIES

Next, it is helpful to note a survey of the control functions and terminal facilities for a high-performance service-type oscilloscope, such as the one illustrated in Fig. 2-7. The meanings of the technical terms in Fig. 2-7 will be clarified in the following pages.

Display of a 60-Hz AC Test Voltage

All oscilloscopes have provision for application of a vertical-input signal. If a 60-Hz test voltage (as from the secondary of a heater transformer) is applied to the VERTICAL INPUT terminals of the oscilloscope, sine-wave patterns such as shown in Fig. 2-8 can be displayed on the screen. When a very slow sweep rate is used (Fig. 2-8D) only a blur is displayed because the successive cycles are highly compressed and cramped. Note in passing that retrace lines are visible in the first two displays; most oscilloscopes (except the least expensive designs) have retrace blanking circuitry. The illustrated patterns may be obtained either with free-running sweep or with trig-gered sweep. When a free-running sweep is used, the pattern will be distorted if a 60-Hz waveform is displayed on 120-, 180-, or 240-Hz sweep, for example—"broken" sections of the pattern will be displayed in a superimposed form. On the other hand, when a triggered sweep is used, a chosen small interval of the waveform (a "window") can be clearly displayed without interference from the remainder of the waveform. This technique will be described subsequently.

Coarse and Fine Deflection Controls

Virtually all oscilloscopes with free-running sweeps have a coarse and a fine (VERNIER) sawtooth sweep frequency control. Most trig-gered-sweep scopes with calibrated time bases also have an uncali-brated vernier sawtooth sweep frequency control. The coarse control is a rotary step switch; the VERNIER control is a potentiometer. Other terms for the same controls are SWEEP RANGE or HORIZONTAL control, and FREQUENCY VERNIER control (see Fig. 2-9). To display a pattern such as that shown in Fig. 2-8, set the HORIZONTAL step control to a position which includes 60 Hz. In Fig. 2-9 this would be the 10–100 position. Then adjust the FREQUENCY VERNIER control to bring the deflection up to nearly 60 Hz, at which point the sync section can obtain control and lock the pattern on the screen. Possibly no other control adjustments will be required, although the GAIN controls will require adjustment if the pattern appears objectionably narrow, wide, high, or low. As shown in Fig. 2-10, the displayed pattern may appear odd unless the VERTICAL and HORIZONTAL GAIN controls are adjusted properly.

Sync-Control Adjustment

At this point it may be found that adjustment of the VERNIER saw-tooth control is comparatively critical in order to maintain a stable pattern on the screen. Thus the pattern may "lock in" for a while, and then "break sync," with display of a blurred area. At the other extreme the displayed pattern might "lock" in very tightly at all set-tings of the VERNIER sawtooth control. In the first case insufficient sync voltage is being used, and the SYNC AMPLITUDE control (Fig. 2-10) should be advanced. In the second case excessive sync voltage is being used; the pattern is being fragmented, and the SYNC AMPLI-TUDE control should be reduced. It is good practice to advance the SYNC AMPLITUDE control only to the point at which solid pattern lock is obtained; this avoids possible pattern distortion due to ex-

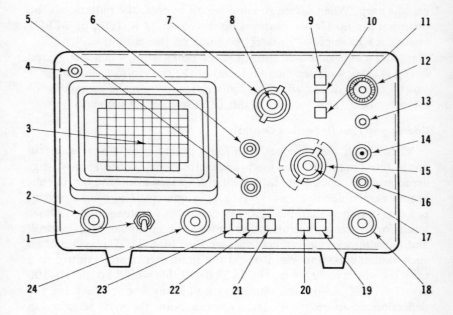

1. **POWER ON-OFF** toggle switch. Applies power to oscilloscope.

2. **INTENSITY** control. Adjusts brightness of trace.

3. **Graticule.** Provides calibration marks for voltage and time measurements.

4. **Pilot lamp.** Lights when power is applied to oscilloscope.

5. ◆▶ **POSITION** control. Rotation adjusts horizontal position of trace. Push-pull switch selects 5X magnification when pulled out; normal when pushed in.

6. ◆ **POSITION** control. Rotation 3adjusts vertical position of trace.

7. **VOLTS/DIV** switch. Vertical attenuator. Coarse adjustment of vertical sensitivity. Vertical sensitivity is calibrated in 11 steps from 0.01 to 20 volts per division when VARIABLE 8 is set to the CAL position.

8. **VARIABLE** control. Vertical attenuator adjustment. Fine control of vertical sensitivity. In the extreme clockwise (CAL) position, the vertical attenuator is calibrated

9. **AC** vertical input selector switch. When this button is pushed in the dc component of the input signal is eliminated.

Fig. 2-7. Survey of control functions and terminal

10. GND vertical input selector switch. When this button is pushed in the input signal path is opened and the vertical amplifier input is grounded. This provides a zero-signal base line. the position of which can be used as a reference when performing dc measurements.

11. DC vertical input selector switch. When this button is pushed in the ac and dc components of the input signal are applied to vertical amplifier.

12. V INPUT jack. Vertical input.

13. $\perp$ terminal. Chassis ground.

14. CAL ⊓ jack. Provides calibrated 0.8 V p-p square wave output at the line frequency for calibration of the vertical amplifier.

15. SWEEP TIME/DIV switch. Horizontal coarse sweep time selector. Selects calibrated sweep times of 0.5 μ SEC/DIV to 0.5 SEC/DIV in 19 steps when VAR/HOR GAIN control 17 is set to CAL. Selects proper sweep time for television composite video waveforms in TVH (television horizontal) and TVV (television vertical) positions. Disables internal sweep generator and displays external horizontal input in EXT position.

16. EXT SYNC/HOR jack. Input terminal for external sync or external horizontal input.

17. VAR/HOR GAIN control. Fine sweep time adjustment (horizontal gain adjustment when SWEEP TIME/DIV switch 15 is in EXT position). In the extreme clockwise position (CAL) the sweep time is calibrated.

18. TRIG LEVEL control. Sync level adjustment determines point on waveform slope where sweep starts. In fully counterclockwise (AUTO) position, sweep is automatically synchronized to the average level of the waveform.

19. TRIGGERING SLOPE switch. Selects sync polarity (+), button pushed in, or (−), button out.

20. TRIGGERING SOURCE switch. When the button is pushed in, INT, the waveform being observed is used as the sync trigger. When the button is out, EXT, the signal applied to the EXT SYNC/HOR jack 16 is used as the sync trigger.

21. TVV SYNC switch. When button is pushed in the scope syncs on the vertical component of composite video.

22. TVH SYNC switch. When button is pushed in the scope syncs on the horizontal component of composite video.

23. NOR SYNC switch. When button is pushed in the scope syncs on a portion of the input waveform. Normal mode of operation.

24. FOCUS control. Adjusts sharpness of trace.

Courtesy B&K Precision Products of Dynascan

facilities for a high-performance oscilloscope.

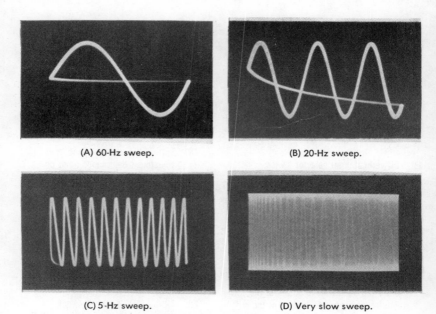

(A) 60-Hz sweep.

(B) 20-Hz sweep.

(C) 5-Hz sweep.

(D) Very slow sweep.

Fig. 2-8. Displays of 60-Hz sine wave at various horizontal-sweep rates.

cessive sync-voltage injection into the sawtooth oscillator. Note that some of the more elaborate oscilloscopes with free-running sweep (such as depicted in Fig. 2-9) have automatic sync action; therefore no SYNC AMPLITUDE control is required.

Some types of waveforms are more difficult to synchronize than others. For example, a narrow pulse waveform may have a large positive excursion and a small negative excursion (see Fig. 2-11). This waveform locks in readily on positive sync, but may be difficult to lock on negative sync. For this reason many service-type oscilloscopes provide a switch for selection of either positive or negative sync. Another type of waveform that may present sync-lock difficulty is the composite video signal. As shown in Fig. 2-12, simple sync systems often provide unstable "lock in" at a 30-Hz deflection rate, although tight "lock in" is provided at a 7875-Hz deflection rate. The difficulty, in this situation, arises from the fact that a simple sync system makes little or no distinction between horizontal-sync tips and vertical-sync tips. For this reason, the better service-type oscilloscopes include horizontal differentiating circuits and vertical integrating circuits in the sync system. Referring back to Fig. 2-7, observe that TVV and TVH sync switches (21 and 22) are provided.

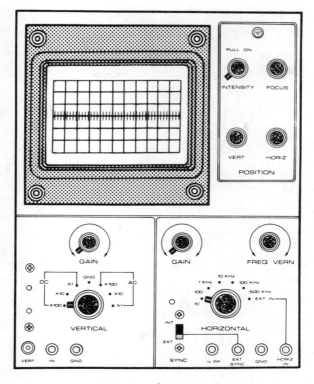

(A) Front-panel arrangement.

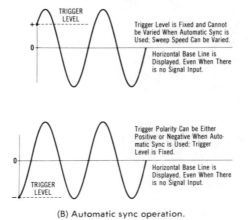

(B) Automatic sync operation.

Fig. 2-9. Operating controls for a modern, general-purpose oscilloscope.

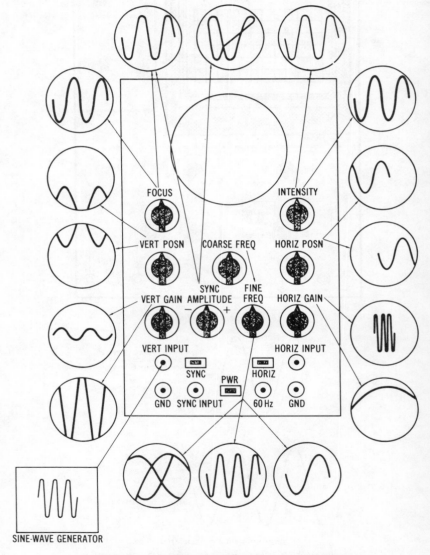

SINE-WAVE GENERATOR

Fig. 2-10. Pattern responses to oscilloscope control adjustments.

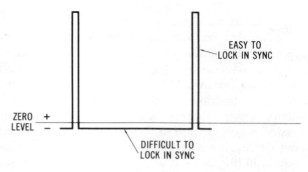

ZERO +
LEVEL −

EASY TO
LOCK IN SYNC

DIFFICULT TO
LOCK IN SYNC

Fig. 2-11. A pulse waveform that has a very small negative excursion.

Intensity-Control Setting vs. Pattern Size

As the VERTICAL GAIN and HORIZONTAL GAIN controls are turned to reduce the size of a sine-wave pattern to a very small area (or a dot), the brightness of the display increases greatly. Conversely, if the GAIN controls are then advanced to expand the sine-wave pattern to full-screen size, the brightness of the display decreases greatly. The reason for this decrease in brightness is that the electron beam

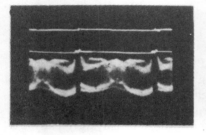

(A) Vertical-rate video signal is often difficult to synchronize.

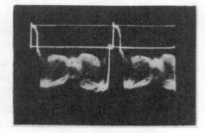

(B) Horizontal-rate video signal is readily synchronized.

Fig. 2-12. "Lock in" with simple sync system.

has a much longer path to trace out; consequently, each point along the trace receives much less energy. It is therefore necessary to turn up the INTENSITY control in order to make a large pattern more clearly visible. Note, however, that there is a practical limit to this increase of intensity; *at a certain upper limit the pattern loses focus greatly and the trace starts to "bloom."* This is the same reaction that occurs in many tv pictures when the BRIGHTNESS control is turned up excessively.

It is evident that if the electron beam travels rapidly over one interval of a waveform, but travels slowly over another interval of the waveform, the brightness of the pattern will vary considerably from one region to another. In Fig. 2-13 the top and bottom edges of the color bar are clearly visible, whereas the central region of the color bar is almost invisible. Similarly, the burst component is displayed at high intensity, whereas the color bar as a whole is displayed at low intensity. Again, the top of the sync pulse is displayed at high intensity, but the leading edge of the pulse is displayed at low intensity. A variation in brightness is unavoidable in display of such complex waveforms.

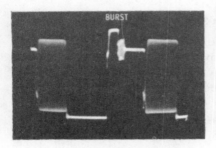

Fig. 2-13. Top and bottom edges of color bar clearly visible but the central region of color bar is almost invisible.

Some oscilloscopes are capable of displaying brighter patterns (in good focus) than other oscilloscopes do. Pattern brightness depends on the value of high voltage applied to the accelerating anode of the crt. If, for example, the accelerating voltage is increased from 1 kV to 2 kV, the available pattern brightness will be greatly increased. A high-performance triggered-sweep oscilloscope typically requires 10 kV for adequate pattern brightness when displaying greatly expanded fast-rise pulses. Accordingly, the phosphor screen can be burned severely if the sweep speed is considerably reduced without the INTENSITY control being turned to a low setting. Increasing the accelerating voltage also increases the energy of the electron beam. In consequence, more energy is required in the vertical- and horizontal-deflection systems in order to move the beam a given distance on the face of the crt. Obviously, higher performance is associated with rapidly increasing production costs. The cost increase is dramatically apparent in oscilloscopes designed for digital troubleshooting of intermittent or unpredictable "glitches." For example, only the highest-performance and expensive oscilloscopes are capable of displaying a 25-nanosecond (25-ns) "glitch" waveform.

Gain-Control Ranges and Settings

As the operator gains familiarity with his or her oscilloscope, he will encounter waveforms that cannot be displayed at satisfactory vertical amplitude, although both the step and vernier GAIN controls have been advanced to their maximum settings. This difficulty, of course, is the result of a low-level vertical-input signal. Although the obvious solution is to obtain another oscilloscope that has higher vertical sensitivity, this is not necessarily the only solution to the problem. For example, it may be feasible to use a direct probe instead of a low-capacitance probe and thereby effectively multiply the vertical sensitivity of the oscilloscope by a factor of 10. A direct probe can be used with an oscilloscope to test low-impedance circuits.

With reference to Fig. 2-14A, the steps of the VERTICAL attenuator "read correctly" when a direct vertical-input cable is used. On the other hand, when a low-capacitance probe is used, each step provides only 1/10 as much vertical gain (Fig. 2-14B). Thus, in effect, the

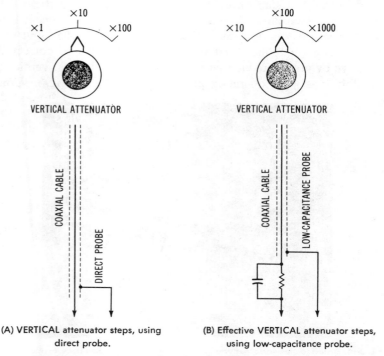

(A) VERTICAL attenuator steps, using direct probe.

(B) Effective VERTICAL attenuator steps, using low-capacitance probe.

Fig. 2-14. Low-capacitance probe reduces vertical gain by a factor of 10.

×1 step becomes a ×10 step when a low-capacitance probe is used. To take an example, the oscilloscope sensitivity might be 20 millivolts per centimeter (20 mV/cm) on the ×1 step of the VERTICAL attenuator when a direct probe is used. Then, if a low-capacitance probe is used, the sensitivity on the ×1 step is only 200 mV/cm. Although the vertical gain is reduced when a low-capacitance probe is utilized, the probe provides an increase of 10 times in the input impedance of the oscilloscope. Therefore a low-capacitance probe permits troubleshooting in high-impedance circuits by minimizing circuit disturbance due to probe loading. If a direct probe were used to check waveforms in high-impedance circuits, the displayed waveforms would be highly distorted or normal circuit action could be completely lost.

Clipping in the vertical amplifier, with resulting pattern distortion, will occur in various service-type oscilloscopes if the operator sets the vernier VERTICAL attenuator to a very low setting and then sets the step VERTICAL attenuator to a very high setting. Even if the input signal is a perfect sine wave, it may appear distorted on the oscilloscope screen (see Fig. 2-15). This type of distortion can occur if the output from the VERTICAL step attenuator overdrives the input stage of the vertical amplifier, and the operator attempts to correct the overdrive by greatly reducing the setting of the VERTICAL vernier attenuator. It is good operating practice to set the VERTICAL vernier

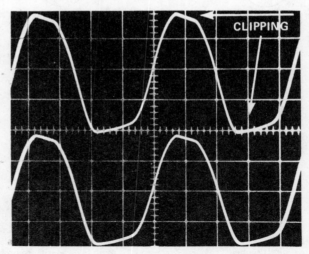

Fig. 2-15. Distorted sine-wave patterns displayed on screen of dual-trace oscilloscope due to improper vertical alternative settings.

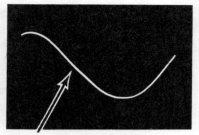

(A) Insufficient horizontal deflection.

(B) Excessive horizontal deflection.

Fig. 2-16. HORIZONTAL GAIN control effect.

attenuator to the midpoint of its range, then to adjust the VERTICAL step attenuator for a reasonable pattern amplitude, and finally to adjust the vertical vernier attenuator for the exact desired amplitude.

Next, the HORIZONTAL GAIN control may need adjustment; the pattern may appear excessively compressed or excessively expanded horizontally, as exemplified in Fig. 2-16. Thus the HORIZONTAL GAIN control(s) must be adjusted. Less elaborate oscilloscopes have a simple potentiometer-type HORIZONTAL GAIN control only; others have both step and continuous GAIN controls. In many service-type oscilloscopes the HORIZONTAL step control is merely a resistive di-

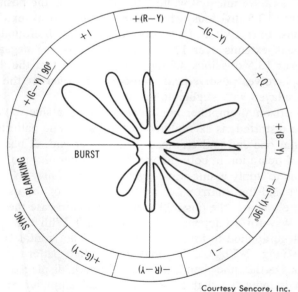

Courtesy Sencore, Inc.

Fig. 2-17. Display of a typical vectorgram.

vider network. Some service-type oscilloscopes, however, have the same high-performance type of compensated step control as provided in the vertical section. This design of service-type oscilloscope is somewhat more expensive; however, it is required for display of vectorgrams (Fig. 2-17) in low-level chroma circuits. If an economy-type oscilloscope is used in this area of troubleshooting, the displayed vectorgrams will be highly distorted. For most test work, however, extended high-frequency response is not required in the horizontal amplifier. If all waveforms are to be displayed on sawtooth sweep, it is only necessary that the horizontal amplifier have sufficient high-frequency response to pass the highest-frequency sawtooth deflection waveform without distortion.

VERTICAL AND HORIZONTAL LINEARITY

Precise troubleshooting procedures require that the oscilloscope have good vertical and horizontal linearity. Otherwise the displayed waveforms will have more or less distortion introduced by the oscilloscope. For example, suppose that a vertical amplifier has 10-percent distortion at full-screen deflection. In such a case a sine wave will be displayed as shown in Fig. 2-18; note that its positive peak and negative peak have unequal widths. In this example the positive peak extends over 13.5 divisions, whereas the negative peak extends over 11 divisions. In this kind of situation it could also be found that the positive peak extends over 11 divisions, whereas the negative peak extends over 13.5 divisions. Observe in Fig. 2-18 that the illustrated distortion involves positive-peak compression, whereas the negative peak is displayed with negligible distortion.

Next, horizontal-amplifier nonlinearity causes a sine-wave pattern to become distorted, as exemplified in Fig. 2-19. The pattern appears expanded at one end and cramped at the other end. A basic test for linearity of operation in both the vertical amplifier and the horizontal amplifier is to apply a sine-wave signal simultaneously to the VERTICAL INPUT and the HORIZONTAL INPUT terminals of the oscilloscope. (This test eliminates the possibility of confusion in case the deflection sawtooth waveform happens to be nonlinear.) With the same sine-wave voltage applied to both the vertical and horizontal channels in the scope (Fig. 2-20A) a diagonal straight-line pattern is normally displayed. On the other hand, if a curved line is displayed, as in Fig. 2-20B, it is indicated that either the vertical amplifier or the horizontal amplifier, or both, must be nonlinear. In the case of a dual-

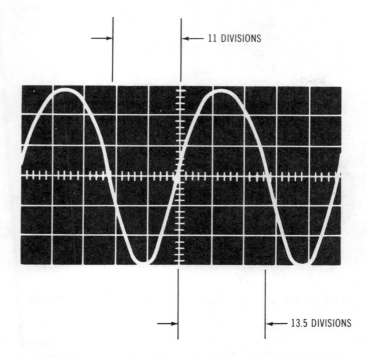

← 11 DIVISIONS

← 13.5 DIVISIONS

Fig. 2-18. Result of vertical-amplifier nonlinearity: positive peak is wider than negative peak.

trace oscilloscope the same sine-wave voltage may be applied to both of the VERTICAL INPUT channels. Then two sine-wave patterns are displayed, as shown in Fig. 2-21A. Next, the POSITION controls are adjusted to superimpose the two traces. If the two traces do not coincide precisely, it means that at least one of the vertical amplifiers is distorting the input signal.

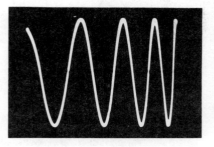

Fig. 2-19. Sine wave distorted by nonlinear horizontal amplifier.

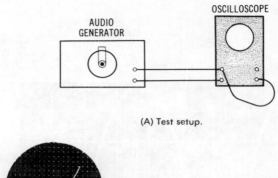

(A) Test setup.

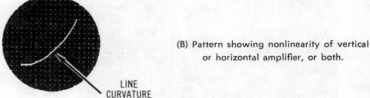

(B) Pattern showing nonlinearity of vertical or horizontal amplifier, or both.

LINE CURVATURE

Fig. 2-20. Linearity test of oscilloscope.

PEAK-TO-PEAK AND PEAK VOLTAGE MEASUREMENTS

Peak-to-peak voltages are specified in receiver service data. An oscilloscope is fundamentally a voltmeter which displays instantaneous, peak, vertical-interval, and peak-to-peak voltages. The meanings of these terms are shown in Fig. 2-22. Note that the zero level is the

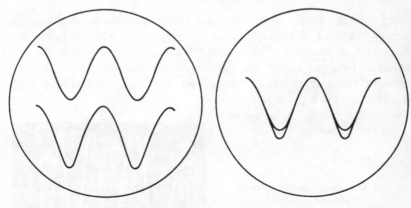

(A) Display of the sense signal applied to both channels.

(B) Superposition of traces is incomplete if one or both amplifiers distort the input signal.

Fig. 2-21. Check of vertical amplifiers in dual-trace oscilloscope.

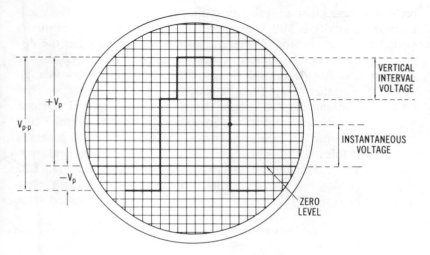

Fig. 2-22. Voltage values measured on an oscilloscope screen.

resting position of the horizontal trace before a vertical-input signal is applied. Suppose that the vertical-deflection sensitivity of the oscilloscope happens to be 1 volt per division. Then, in the example of Fig. 2-22, the peak-to-peak voltage ($V_{\text{p-p}}$) is 16 volts; the positive-peak voltage ($+V_{\text{p}}$) is 13 volts; the negative-peak voltage ($-V_{\text{p}}$) is 3 volts; the vertical-interval voltage is 5 volts; the instantaneous voltage is +5 volts. Observe that all of these voltages are measured in the same unit: 1 volt per vertical division. Most waveforms have both positive-peak and negative-peak voltages. As noted previously, however, a waveform may be all positive, or it may be all negative.

Modern oscilloscopes generally have highly stable and inherently calibrated vertical amplifiers. With reference to Fig. 2-23 the VERTICAL step attenuator indicates peak-to-peak voltage per vertical division. In this example the oscilloscope sensitivity is 0.05 V p-p per

Fig. 2-23. Example of a VERTICAL step attenuator that indicates peak-to-peak voltage per vertical division.

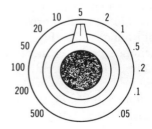

division on the highest sensitivity setting of the attenuator. The oscilloscope sensitivity is 500 V p-p per division on the lowest sensitivity setting of the attenuator. Thirteen steps are provided so that the displayed pattern can be adjusted to desired height without recourse to adjustment of the vernier VERTICAL GAIN control. In other words, although the vernier control can be used to "fill in" between positions on the step attenuator, the vernier control is uncalibrated. Therefore, when peak-to-peak voltage values are to be measured, the vernier control must be turned to its maximum position (snap detent is generally provided at this maximum position). Note that the peak-to-peak voltage values indicated by the VERTICAL step attenuator in Fig. 2-23 apply to oscilloscope operation with a direct probe. If a low-capacitance probe is used, the operator must multiply each indicated value by 10.

Obsolescent oscilloscopes require calibration each time that the vernier GAIN control is reset; the GAIN control must be frequently reset because only three or four positions are provided on the step attenuator. To calibrate an old-model oscilloscope for peak-to-peak voltage measurements, a known peak-to-peak voltage is applied to its VERTICAL INPUT terminals, and the resulting number of divisions is noted for deflection along the vertical axis. For example, if a 1-volt peak-to-peak signal is applied to the oscilloscope and 10 divisions of vertical deflection are observed, it follows that the VERTICAL GAIN controls are set for a sensitivity of 0.1 volt per division. If the calibration determination is made using a low-capacitance probe, and changeover is subsequently made to a direct probe, the sensitivity factor must then be divided by 10.

Virtually all modern oscilloscopes have provision for applying a built-in square-wave voltage to the input of the vertical amplifier when desired. This feature permits a quick check of the rated calibration factor in case of doubt; the square-wave voltage has a specified value, such as 1 V p-p, and often has a frequency of 1 kHz. Although some other frequency, such as 60 Hz, could be used, it is advantageous to employ a higher frequency, such as 1 kHz, because a 1-kHz square wave also provides a useful check of low-capacitance probe adjustment. (See Fig. 2-24.) If the trimmer capacitor in a low-capacitance probe is incorrectly adjusted, displayed waveforms will be more or less differentiated or integrated. This topic of distortion introduced by probes will be explained in greater detail subsequently. Older models of oscilloscopes often provided a front-panel binding post with a 1-V p-p, 60-Hz, sine-wave source for vertical-

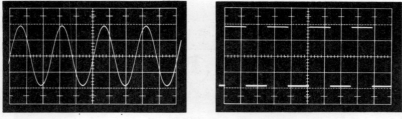

(A) Sine-wave and square-wave calibration voltages.

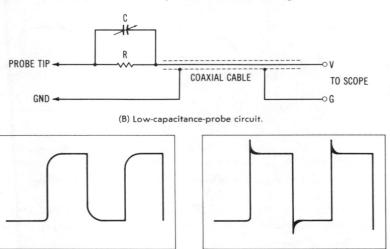

(B) Low-capacitance-probe circuit.

UNDERCOMPENSATED PROBE   OVERCOMPENSATED PROBE

(C) 1-kHz square-wave calibration also checks low-capacitance-probe response.

Fig. 2-24. Built-in calibration voltages may have a sine or square waveform.

amplifier calibration. An occasional old-model oscilloscope provided a built-in 1-V p-p, 60-Hz, square-wave calibrating source.

DC VOLTAGE COMPONENTS AND MEASUREMENTS

State-of-the-art oscilloscopes provide a choice of direct-coupled or ac-coupled VERTICAL INPUT (dc response or ac response). A dc scope has a low-frequency response down to zero frequency, or dc. An ac scope, however, has some definite low-frequency limit, such as 20 Hz. The typical response of a dc oscilloscope to battery voltages is illustrated in Fig. 2-25. If a 10-volt battery, for example, is connected to the VERTICAL INPUT terminals of a dc scope, a positive polarity deflects the beam upward, and the beam remains deflected

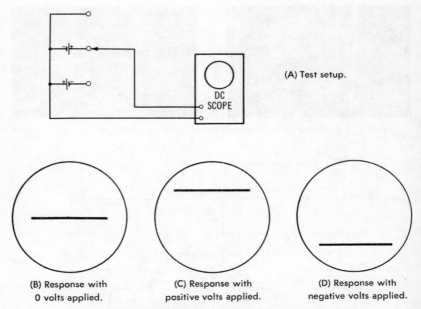

(A) Test setup.

(B) Response with
0 volts applied.

(C) Response with
positive volts applied.

(D) Response with
negative volts applied.

Fig. 2-25. Application of battery voltages to a dc oscilloscope.

until the dc voltage is removed. Similarly, when the terminal polarity is reversed, the beam is deflected downward from its resting position by the same amount.

Next, if the 10-volt battery is connected to the input terminals of an ac oscilloscope, a positive polarity deflects the beam upward for a brief instant. Then the beam returns to its original resting position although the battery is still connected to the vertical-input terminals of the oscilloscope. Now, if the battery is removed and the VERTICAL INPUT terminals of the oscilloscope are shorted together, the beam is deflected downward for a brief instant. Then the beam returns to its original resting position. Note that the beam is deflected downward because the series coupling capacitor in the vertical-input circuit was previously charged; then, when the VERTICAL INPUT terminals are shorted, the coupling capacitor discharges and passes a transient voltage into the vertical amplifier.

It follows that if the 10-volt battery is connected to the VERTICAL INPUT terminals of an ac oscilloscope, a negative polarity will deflect the beam downward for a brief instant. Then the beam will return to its original resting position, although the battery is still connected to the VERTICAL INPUT terminals. Now, if the battery is removed and the VERTICAL INPUT terminals of the oscilloscope are shorted together, the beam will be deflected upward for a brief instant. In other words, the vertical-input coupling capacitor discharges and passes a transient exponential voltage into the vertical amplifier. Note in passing that a dc oscilloscope can be calibrated with a dc voltage source, and the sensitivity of the vertical amplifier will be the same as if the calibration were made with a sine wave in terms of peak-to-peak voltage.

AC WAVEFORMS WITH DC COMPONENTS

Many of the waveforms encountered in troubleshooting with the oscilloscope have dc components. Consider a simple transistor amplifier stage, as shown in Fig. 2-26. It is evident that the output sine waveform "rides on" V_D, or, the output waveform has a dc component equal to V_D. This dc component will be rejected by an oscilloscope operating on its ac input function, but the dc component will displace the sine wave upward on the screen if the oscilloscope is operated on its dc input function. All oscilloscopes that have dc response are provided with a switch for changing from dc to ac response. This changeover is accomplished by switching a series coupling capacitor in series with the vertical-input lead. When ac response is used, all waveforms will be displayed centered on the resting level of the horizontal trace. However, when dc response is

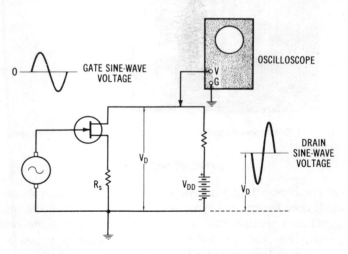

Fig. 2-26. The output sine wave has a V_{DD} dc component.

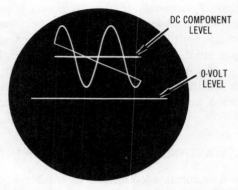

Fig. 2-27. Response of a dc oscilloscope to ac voltage with a dc component.

employed, a waveform with a dc component will be displaced vertically with respect to the resting level. It will be centered on the dc component level, as shown in Fig. 2-27. With reference to Fig. 2-28, the composite video signal output from a picture detector has a single-polarity excursion with a dc component as indicated.

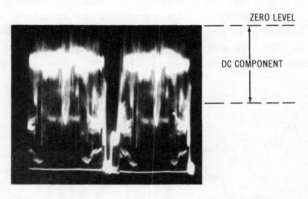

Fig. 2-28. Output waveform from the picture detector has a dc component.

SYNC FUNCTION

Most troubleshooting with the oscilloscope is done with patterns that are locked by internal sync. As shown in Fig. 2-29 the internal-sync function operates from a sample of the input signal voltage. Some tests, however, cannot be made satisfactorily unless a sync voltage separate from the signal voltage is used; in such a case the external-sync function must be utilized. A basic example concerns checking of delay-line operation in a color receiver with a simple oscilloscope. Referring to Fig. 2-30, the horizontal-sync pulse has a

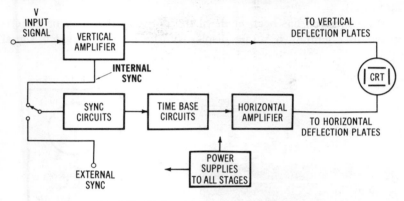

Fig. 2-29. Chief sections in a basic oscilloscope.

width of 4.75 μs, and the pedestal has a width of 11 μs. After passage through the delay line, the pulse is normally delayed approximately 1 μs. Subnormal, abnormal, or no delay causes poor "color

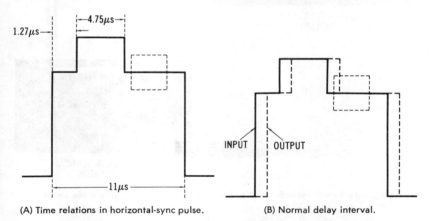

(A) Time relations in horizontal-sync pulse.

(B) Normal delay interval.

Fig. 2-30. Normal action of delay line in Y amplifier of color receiver.

fit." To check the delay-line action the external-sync function of the oscilloscope must be used. For example, the input sync-pulse waveform may be applied to the EXTERNAL SYNC terminal of the oscilloscope. Then the vertical input lead of the oscilloscope is applied first at the input of the delay line, and then at the output of the delay line. If the second pattern is displaced with respect to the first pattern, as depicted in Fig. 2-30B, it is indicated that the delay-line action is normal.

Prior mention has been made of the difficulty that may be experienced in obtaining a stable display of the composite video signal

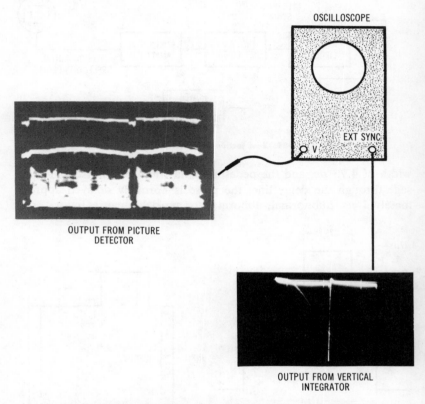

OUTPUT FROM PICTURE
DETECTOR

OUTPUT FROM VERTICAL
INTEGRATOR

Fig. 2-31. External-sync function provides stable lock of composite video signal.

at a 30-Hz deflection rate with a simple service-type oscilloscope. Note that use of the external-sync function can provide stable pattern display in this situation, as shown in Fig. 2-31. Here the output waveform from the picture detector is fed to the VERTICAL INPUT terminal of the oscilloscope, and the pattern is locked by means of the pulse output from the vertical integrator.

ACTION OF TRIGGERED-SWEEP CONTROLS

A free-running time base is self-oscillatory and generates a continuous sawtooth output whether or not an input signal is applied to the vertical amplifier. On the other hand, a triggered time base is not self-oscillatory; it generates a sawtooth output only when the leading edge of a signal voltage is applied to the vertical amplifier. Then, a single sawtooth excursion is generated, and the time-base oscillator stops; if another leading edge is applied to the vertical amplifier, another sawtooth excursion will be generated. Most modern service-type oscilloscopes provide a choice of free-running (recurrent) sweep or triggered sweep, and are termed *dual-mode* oscilloscopes.

Thus one advantage of a triggered-sweep scope is that the sweep starting time is under operator control; as depicted in Fig. 2-32 the color burst can be picked out of the video waveform and expanded on the screen of a triggered-sweep oscilloscope. With few exceptions any small portion of any waveform can be selected for display, and greatly expanded on the screen of a triggered-sweep oscilloscope. Virtually all triggered time bases are calibrated, and the sweep speed is indicated by the setting of the TIME BASE control, as shown in Fig. 2-33. In this example the control has a triggered-sweep range from 0.1 microsecond per division to 100 milliseconds per division. This range permits the operator to measure frequency through the audio and rf range, and to measure pulse widths, rise time, settling time, slew rate, and associated waveform characteristics that involve time. In Fig. 2-33 the inner knob is a vernier adjustment of sweep speed, and it is uncalibrated. Accordingly when one is reading sweep-speed values from the scale, it is necessary to turn the vernier control fully clockwise.

Trigger Level and Slope Control

All but the economy-type triggered-sweep oscilloscopes have a TRIGGER LEVEL control and a SLOPE control, as exemplified in Fig. 2-34. The operator can select the point along the leading edge or along the trailing edge at which the time base will be triggered—this trigger point determines the start of the display. Observe in Fig. 2-34 that the SLOPE switch provides a choice of positive triggering or negative triggering. Thus in Fig. 2-34A the sine wave is being triggered on its positive rising slope in the top diagram; in the bottom diagram the SLOPE control is set to its negative position, with the result that the sine wave is triggered on its falling part. This is a point on the

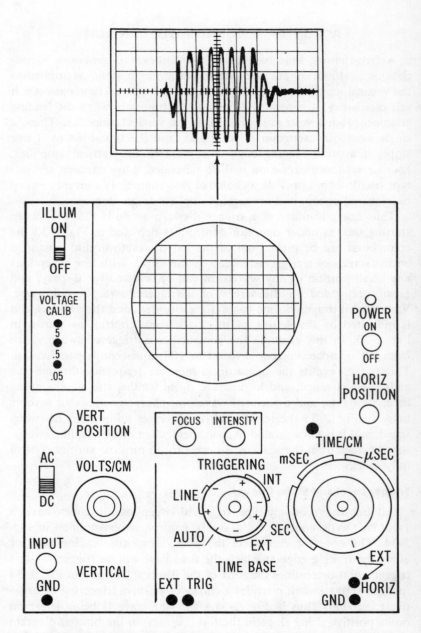

Fig. 2-32. Color burst can be picked out of the video waveform and expanded on the screen of a triggered-sweep oscilloscope.

positive portion of the falling interval *because the* TRIGGER LEVEL *control is set to the positive portion of its range.* Next, in Fig. 2-34B, the waveform is being triggered on its positive rising slope in the top diagram; in the bottom diagram the SLOPE control is set to its negative position, with the result that the sine wave is triggered on its falling part. This is a point on the negative portion of the falling part *because the* TRIGGER LEVEL *control is set to the negative portion of its range.*

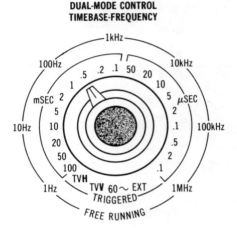

DUAL-MODE CONTROL
TIMEBASE-FREQUENCY

○ TIME PER DIV–FREQ
● TURN FULL CLOCKWISE TO READ SWEEP SPEED

TVH -PRESET 7875-Hz SWEEP
TVV -PRESET 30-Hz SWEEP

Fig. 2-33. Typical calibrated TIME BASE control for a triggered-sweep oscilloscope.

Observe the waveform in the top diagram of Fig. 2-34A. It is being triggered near its positive-peak value *because the* TRIGGER LEVEL *control is set to a comparatively high point in its positive range.* Now, if the operator backs off on the setting of the TRIGGER LEVEL control, to a comparatively low point in its positive range, *the sine wave will then be triggered near its 0-volt value.* Similarly, with respect to the waveform in the top diagram of Fig. 2-34B, the sine wave is being triggered near its 0-volt value *because the* TRIGGER LEVEL *control is set to a comparatively high point in its negative range.* Now, if the operator backs off on the setting of the TRIGGER LEVEL control, to a comparatively low point in its negative range, *the sine*

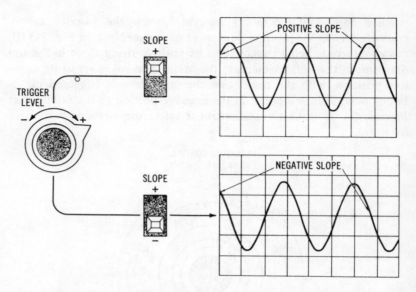

(A) Sine wave of top diagram being triggered at a high positive level on its positive slope; sine wave of bottom diagram at a low positive level on its negative slope.

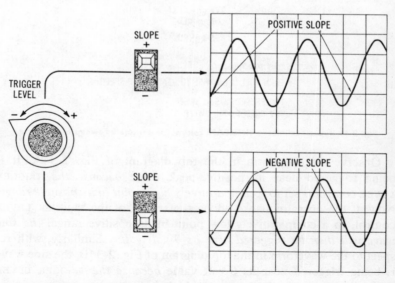

(B) Sine wave of top diagram being triggered at a low negative level on its positive slope; sine wave of bottom diagram at a high negative level on its negative slope.

Fig. 2-34. TRIGGER LEVEL control and SLOPE control actions.

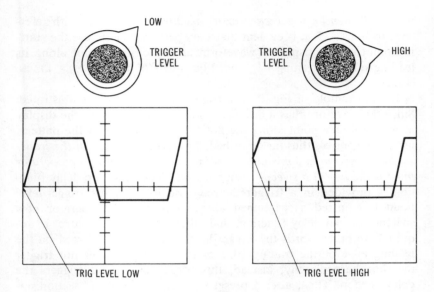

(A) This waveform is being triggered along its positive slope (rising interval).

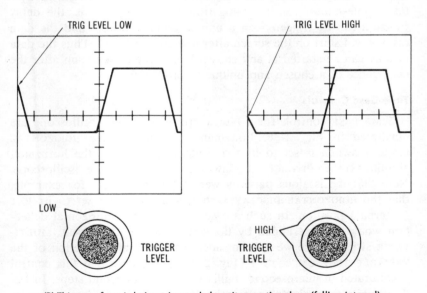

(B) This waveform is being triggered along its negative slope (falling interval).

Fig. 2-35. The starting point of the displayed waveform can be chosen at any desired point on the leading or trailing edge by adjustment of the TRIGGER LEVEL control.

wave will then be triggered near its negative-peak value. With reference to Fig. 2-35, it is evident that the operator can choose the starting point of the displayed waveform at any desired point along its leading edge or trailing edge by adjustment of the TRIGGER LEVEL control.

In the example of Fig. 2-35 a trapezoidal waveform is illustrated. Since the waveform has a flat top it is not possible to start the display at any arbitrary point along the flat top (or bottom) of the pattern, using the controls thus far described. *To start the display of the trapezoidal waveform at some chosen point along its flat top, the operator must make use of a sweep-delay function.* When a sweep-delay function is provided in a triggered-sweep oscilloscope, the trigger-level point is adjusted as explained above; however, the operator now switches on the delay function, and adjusts the DELAY control as required. In other words the trigger-level point that is selected on the leading edge or the trailing edge of the waveform does not trigger the time base directly; instead, this trigger-level point triggers the delay section. Then, after a preset time delay, the delay section applies an output pulse to the time base of the oscilloscope, and the display starts on the screen. The delay function is as useful in digital troubleshooting as in analog troubleshooting; that is, the delay function can be triggered on a unique digital "word," and the data display will start on the screen after a waiting interval. Thus the data display can be started at any chosen line in the data stream after the occurrence of a chosen and unique "word" trigger.

Time-Base Controls

TIME BASE controls for a typical triggered-sweep oscilloscope are illustrated in Fig. 2-36. For many applications the HORIZONTAL DISPLAY switch is set to its INT (internal) position; the horizontal amplifier is then driven by the sawtooth generator of the oscilloscope. Note that if Lissajous patterns were to be displayed, for example, that the HORIZONTAL DISPLAY switch would then be set to its EXT (external) position. In such a case the amount of horizontal deflection would be determined by the setting of the EXT LEVEL MULTIPLIER switch. Here we will consider the "internal" operation of the VARIABLE TIME CM control (Fig. 2-36C). Observe that this control is calibrated in microsecond, millisecond, and 1-second steps. In addition, a VAR (variable) setting is also provided—the sweep time is uncalibrated when operating in this position. In most troubleshooting procedures the TIME BASE is operated on one of its calibrated set-

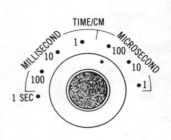

(A) SPEEP RATE control set to 100 μs/cm.

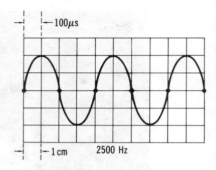

(B) Display of 2500-Hz sine wave.

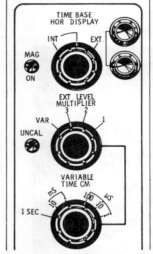

(C) Typical group of TIME BASE controls.

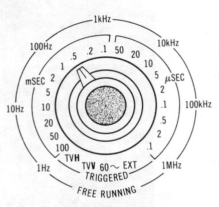

(D) An elaborate time-per-centimeter control with 19 steps.

Fig. 2-36. TIME BASE controls for a triggered-sweep oscilloscope.

tings so that the operator can measure rise time, delay time, frequency, pulse width, and so on.

Let us observe how a waveform can be expanded for analysis of detail by operating the time base at high speed. In Fig. 2-37A a display of the sync pulse and color burst is illustrated for an economy-type service oscilloscope. In Fig. 2-37B an expansion of the sync pulse and color burst on triggered sweep with dual-trace display is exemplified. The sweep speed is set to 4 μs per division. Each cycle in the burst is clearly visible. Observe that the sync pulse occupies

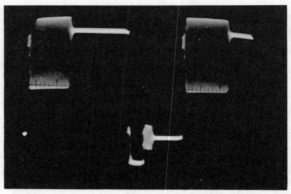

(A) Pattern produced by economy-type service oscilloscope.

←— DIVISION —→

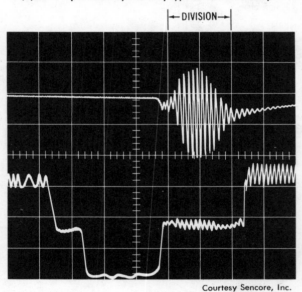

Courtesy Sencore, Inc.

(B) Dual-trace pattern provided by triggered-sweep at 4 μs/division sweep speed.

Fig. 2-37. Display of sync pulse and color burst.

approximately 4.75 μs at its base, and that approximately 14 cycles of burst appear within one division. What is the significance of the wavy top in the horizontal-sync pulse? This is an indication of the presence of residual 920-kHz interference, due to visible beating between the 3.58-MHz color subcarrier and the 4.5-MHz sound signal. When the 920-kHz beat interference is substantial, it appears

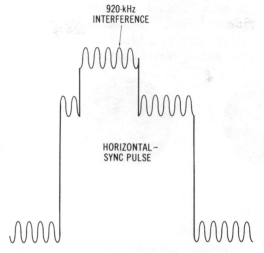

Fig. 2-38. Appearance of 920-Hz interference in the video signal.

as shown in Fig. 2-38, and it produces appreciable interference in the color picture.

Trigger Controls

Trigger controls for a triggered-sweep oscilloscope are shown in Fig. 2-39. Polarity switching permits triggering on either the positive or the negative excursion of a waveform. Note that most waveforms are displayed in the ac trigger mode. To the beginner the DC TRIGGER MODE control may be misleading—in this case, the term "DC" denotes merely that low-frequency signals only are permitted to pass into the trigger section of the oscilloscope. This is a useful feature for providing stable display and expansion of the color burst, for example (any tendency to high-frequency "jitter" is rejected). Note that in the AUTO (automatic) trigger mode, triggering action is essentially the same as in a free-running time base—however, synchronization is automatic and no SYNC AMPLITUDE control is utilized. An advantage of this mode of operation is that a horizontal line is displayed even though no vertical-input signal is applied—a helpful feature for the beginning operator. A disadvantage of this mode of operation is that there is no control of the point along a waveform at which triggering occurs.

Observe the STABILITY and TRIGGER LEVEL dual control in Fig. 2-39D. The operation of the TRIGGER LEVEL control has been explained. The STABILITY control does not require frequent attention. To set the STABILITY control, the oscilloscope must be operated in its NORM (normal triggering) mode. With the TRIGGER LEVEL control set to one extreme end of its range, the STABILITY control is advanced

(A) Pattern with STABIL-ITY control set to extreme right-hand end of its range.

(B) Pattern with STABIL-ITY control set to mid-point of its range.

(C) Pattern with STABIL-ITY control set to extreme left-hand end of its range.

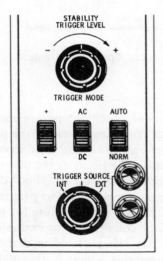

(D) Typical group of trigger controls.

Fig. 2-39. Trigger controls and patterns.

clockwise until a pattern is displayed on the screen. Then the STABILITY control is backed off slightly, so that the pattern just disappears. This is the correct operating position for the STABILITY control, and it will seldom require attention. To become familiar with triggered-sweep operation the apprentice should start with its simplest mode: AUTO triggering. Then, after he or she learns the response of the various controls associated with free-running sweep, he can switch

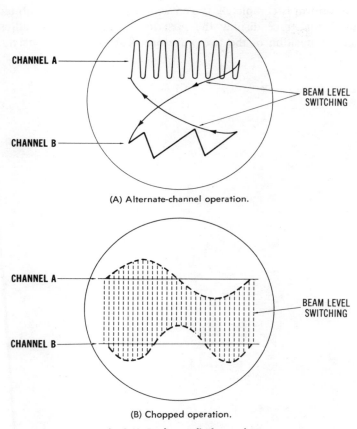

(A) Alternate-channel operation.

(B) Chopped operation.

Fig. 2-40. Dual-trace display modes.

over to the NORM mode and learn the responses provided by the remaining time-base controls.

DUAL-TRACE DISPLAY MODES

Most dual-trace oscilloscopes provide a choice of alternate-channel operation, or chopped operation, as shown in Fig. 2-40. When low-frequency waveforms are to be displayed, chopped operation is desirable to avoid objectionable "flicker." The chief disadvantage of this mode is that each waveform is displayed as a succession of dots or dashes. When medium-frequency or high-frequency waveforms are to be displayed, alternate-channel operation is preferred. In this mode each waveform is completely traced without interruption. With either mode of dual-trace display the POSITION controls can be adjusted for superimposition of the two waveforms, as noted previously.

CHAPTER 3

Using Oscilloscope Probes

Various kinds of oscilloscope probes are required in particular types of tests and measurements. But, it should not be supposed that probes are *always* required. For example, consider the power-supply ripple check depicted in Fig. 3-1. The output of a power supply

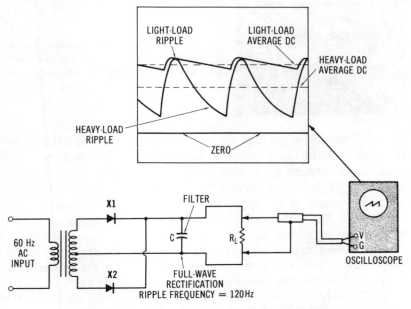

Fig. 3-1. Direct probe may be used to check power-supply ripple.

represents a *low-impedance circuit*. In turn, *the input impedance of the oscilloscope cannot load the circuit significantly.* This means that we can use a direct probe with coaxial cable, or even a pair of open test leads to feed the ripple voltage to the VERTICAL INPUT terminals of the oscilloscope. Receiver service data may specify the maximum tolerable peak-to-peak ripple voltage. The reason why open test leads may be utilized to check ripple voltage is that the low impedance of the power-supply output circuit makes the test leads immune to pickup of stray fields, which would otherwise interfere with the displayed pattern.

WHY PROBES ARE NEEDED IN OSCILLOSCOPE TESTS AND MEASUREMENTS

By way of example, suppose that we have a pair of open test leads connected to the VERTICAL INPUT terminals of an oscilloscope, and that we leave the test leads lying on the bench. A 60-Hz pattern will be displayed on the oscilloscope screen, unless the VERTICAL GAIN controls are set to a low value. (See Fig. 3-2.) In other words, a

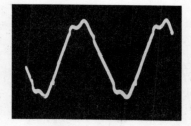

Fig. 3-2. Example of stray-field voltage picked up by open test lead to the VERTICAL INPUT terminal of an oscilloscope.

"floating" vertical-input lead picks up stray fields because it is capacitively coupled to the power-line wiring in the wall. Although the capacitance between a test lead and a power wire several feet away is very small, it is significant in this situation because the vertical-input impedance of the oscilloscope is very high. This small stray capacitance has a decreasing reactance at higher frequencies and tends to act as a high-pass filter. Thus the higher-frequency power-line harmonics appear more prominent in the display than if the vertical-input test lead were directly connected to the power line.

An oscilloscope has some input capacitance at its VERTICAL INPUT terminal, such as 20 or 30 pF, as noted in Fig. 3-3; a pair of test leads has a capacitance from 5 to 50 pF, depending upon the separa-

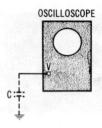

OSCILLOSCOPE

(A) VERTICAL INPUT terminal has a
capacitance of 20 to 30 pF.

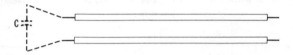

(B) A pair of test leads has a capacitance from 5 to 50 pF.

(C) A coaxial input cable has a capacitance from 50 to 80 pF.

Fig. 3-3. Capacitances in the input circuit of an oscilloscope.

tion of the leads; a coaxial input cable has a capacitance from 50 to 80 pF. Troubleshooting with the oscilloscope requires connection of test leads or a coaxial cable to the vertical-input connector of the oscilloscope. Unshielded test leads are often unsuitable for testing in tv circuitry or audio circuitry. FET circuits often have high impedance, and unshielded vertical-input leads to the oscilloscope will pick up excessive flyback-pulse and hum interference. It is standard practice, therefore, to make all tv and audio waveform tests with a coaxial input cable connected to the vertical-input connector of the oscilloscope.

It follows from Fig. 3-3 that the total input capacitance of the vertical-input terminal with a coaxial cable will be in the range from 75 to 100 pF. This arrangement is termed *direct-cable input*. A direct cable can be used to check the waveform across an emitter resistor without appreciable waveform distortion. On the other hand, a direct cable can impose excessive capacitive loading at the gate or drain of an FET, for example. An example of substantial capacitive loading on waveform display is illustrated in Fig. 3-4. Therefore it is

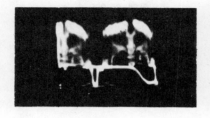 

(A) Display of a normal video signal.　　(B) Signal distorted by integration.

Fig. 3-4. Example of waveform distortion due to excessive capacitive loading of a high-impedance circuit.

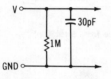

(A) General equivalent vertical-input circuit of typical oscilloscope.

(B) General equivalent input circuit with coaxial input cable.

(C) Equivalent input circuit at low frequencies.

(D) Equivalent input circuit at high frequencies.

Fig. 3-5. Oscilloscope input parameters.

general practice to use a low-capacitance probe with an oscilloscope. Note that a low-capacitance probe (when properly adjusted) does not change the low-frequency response or the high-frequency response of the oscilloscope. The probe merely reduces the effective sensitivity of the oscilloscope by a factor of 10.

LOW-CAPACITANCE-PROBE CONSTRUCTION AND ADJUSTMENT

A general equivalent vertical-input circuit for an oscilloscope is shown in Fig. 3-5A. The equivalent circuit comprises 1 megohm of resistance shunted by 30 pF of capacitance. Next, when a coaxial input cable is connected to the oscilloscope, its general equivalent input circuit becomes 1 megohm of resistance shunted by 100 pF of capacitance, as exemplified in Fig. 3-5B. Now, consider how this equivalent circuit "looks" to a low-frequency voltage waveform. Since the reactance of the shunt capacitance is extremely high at very low frequencies, the vertical-input circuit will "look" resistive, as depicted in Fig. 3-5C. Next, consider how the vertical-input equivalent circuit "looks" to a high-frequency voltage waveform. Inasmuch as the re-

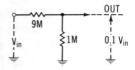

(A) At low frequencies 9-MΩ series resistor divides input signal voltage by 10.

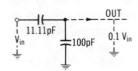

(B) At high frequencies 11.11-pF series capacitor divides the input signal voltage by 10.

(C) At any frequency 9-MΩ resistor in shunt with 11.11-pF capacitor divide input signal voltage by 10.

LOW-CAPACITANCE PROBE

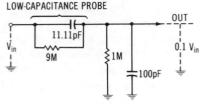

(D) External appearance of typical probe.

Fig. 3-6. Principles of low-capacitance probe.

actance of the shunt capacitance will be quite low at very high frequencies, the vertical-input circuit will "look" capacitive, as depicted in Fig. 3-5D. With these vertical-input parameters in mind, we will consider how an *RC* probe can be devised which will reduce the effective input capacitance of the oscilloscope by a factor of 10, and without incurring any waveform distortion.

With reference to Fig. 3-6A, for low-frequency operation the effective input resistance of the oscilloscope will be increased by a factor of 10 if a 9-megohm resistor is connected in series with the vertical-input lead. Next, with reference to Fig. 3-6B, for high-frequency operation the effective input capacitance of the oscilloscope will be decreased by a factor of 10 if an 11.11-pF capacitor is connected in series with the vertical-input lead. In turn, it is reasonable to suppose that at any frequency of operation the effective input impedance of the scope would be increased by a factor of 10 if a low-capacitance probe employed a 9-megohm resistor and an 11.11-pF capacitor, as shown in Fig. 3-6C. This is a fact that can be proved both experimentally and mathematically. The arrangement in Fig. 3-6A is called *resistive voltage division,* and the arrangement in Fig. 3-6B is called *capacitive voltage division*. Observe that the time constant (RC product) of the low-capacitance probe in Fig. 3-6C is 99.99×10^{-6}, and that the time constant of the oscilloscope input circuit is 100×10^{-6}. It is this equality of time constants for probe and oscilloscope that provides distortionless display of waveforms.

Note in Fig. 3-6C that the low-capacitance probe has the same attenuation factor for dc as for ac voltages. In other words, the low-capacitance probe does not disturb the normal response of a dc oscilloscope. Next, a word of caution: Not all oscilloscopes have 1-megohm input resistance and 30 pF (or 100 pF with coaxial cable) input capacitance. Therefore the R and C values in a low-capacitance probe must be correctly selected for the oscilloscope with which the probe will be used. Virtually all low-capacitance probes provide an adjustment of series capacitance, although few provide any adjustment of series resistance. Capacitance values are comparatively critical with regard to waveform distortion. Note also that there is a practical consideration involved in a design choice of the attenuation factor of 10—if a direct probe is substituted for a low-capacitance probe, the operator simply moves the decimal point in his calibration value one place to the left. This is a simple process, compared with an arbitrary arithmetical calculation.

The adjustment of the trimmer in a low-capacitance probe can be checked as shown in Fig. 3-7. A 10-kHz sine-wave voltage from a generator is applied to an oscilloscope via a direct probe, and then via a low-capacitance probe. The first check is made on the "10" step of the VERTICAL attenuator, and the second check is made on the "1" step of the VERTICAL attenuator. If the low-capacitance probe is in correct adjustment, both displays will appear with equal height on the screen. On the other hand, if a disparity is observed in displays from the two probes, the trimmer in the low-capacitance probe

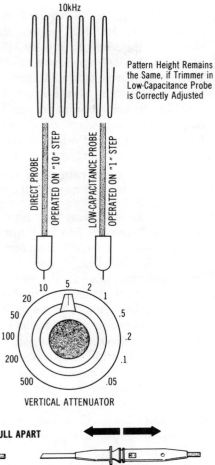

10kHz

Pattern Height Remains
the Same, if Trimmer in
Low-Capacitance Probe
is Correctly Adjusted

DIRECT PROBE
OPERATED ON "10" STEP

LOW-CAPACITANCE PROBE
OPERATED ON "1" STEP

(A) Checking adjustment of trimmer in low-
capacitance probe.

10   5   2
20   1
50   .5
100   .2
200   .1
500   .05

VERTICAL ATTENUATOR

1. PULL APART

PROBE COMPENSATION
ADJUSTMENT

CLP-18 TIP

2. ROTATE 180°

3. PUSH BACK TOGETHER

(B) Combination direct/low-capacitance probes.

Courtesy B&K Precision Products of Dynascan

Fig. 3-7. Direct and low-capacitance probes.

should be adjusted as required. Note that when the test frequency is reduced to 100 Hz, the response of the low-capacitance probe should be the same as at a 10-kHz test frequency. In case that the height of the 100-Hz pattern is different from the height of the 10-kHz pattern, it is indicated that the resistor in the low-capacitance probe is off value. Also, a word of caution: A good audio oscillator provides the same output voltage at 100 Hz as at 10 kHz. An economy-type oscillator, however, may lack uniform output; this possibility can be checked using the direct probe.

CONSTRUCTION AND OPERATION OF DEMODULATOR PROBES

Demodulator systems may employ either series detectors or shunt detectors. Both types find application in demodulator probes. Troubleshooting with the oscilloscope may involve tests in circuits operating at 20 MHz, 40 MHz, or even higher frequencies. However, service-type oscilloscopes seldom have vertical-amplifier frequency response beyond 5 MHz. Therefore, to display waveforms in high-frequency circuitry a demodulator probe must be used (a low-capacitance probe has no response in this situation). A demodulator probe is a simple detector arrangement that operates on the same basic principle as the picture detector in a tv receiver; a diode rectifier and associated circuit recover the modulation envelope from a high-frequency amplitude-modulated carrier. Note that the modulation envelope contains low frequencies which are within the response range of the vertical amplifier of the oscilloscope. Basic series and shunt detector configurations are shown in Fig. 3-8; the shunt arrangement has a comparatively high output impedance.

The simplest demodulator-probe design is shown in Fig. 3-9; it consists of a semiconductor diode connected in series with a coaxial cable. As depicted in Fig. 3-9C the cable capacitance serves as a charging capacitor for the diode—it operates as a low-pass filter. The dc charge that builds up on the capacitor proceeds to discharge through the 1-megohm resistance of the VERTICAL attenuator and also through the back (reverse) resistance of the diode. This probe has limited usefulness at very high frequencies because the coaxial cable does not "look" like a simple capacitor at higher frequencies. The cable develops standing waves; in consequence it has a very high input impedance at one resonant frequency, and has a very low input impedance at another of its resonant frequencies. Therefore more elab-

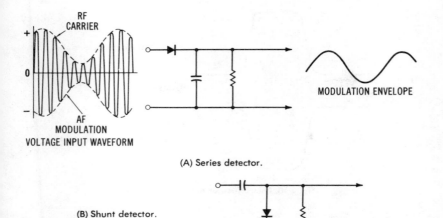

(A) Series detector.

(B) Shunt detector.

Fig. 3-8. Basic detector arrangements.

orate probe circuitry is preferred in practical troubleshooting procedures. Note that if the diode is polarized one way, a negative-going output is obtained; if the diode is polarized the other way, a positive-going output is obtained, as shown in Fig. 3-10.

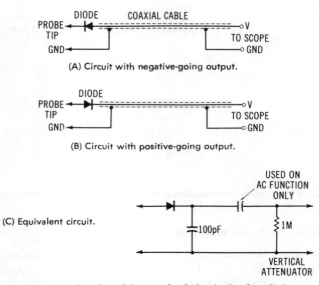

(A) Circuit with negative-going output.

(B) Circuit with positive-going output.

(C) Equivalent circuit.

Fig. 3-9. Simplest demodulator probe design (series detection).

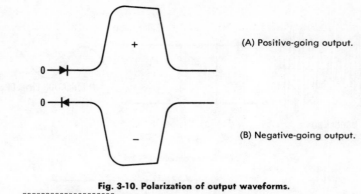

(A) Positive-going output.

(B) Negative-going output.

Fig. 3-10. Polarization of output waveforms.

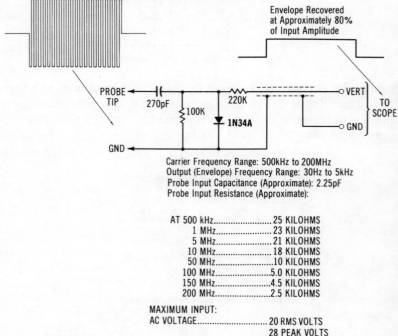

Envelope Recovered
at Approximately 80%
of Input Amplitude

Carrier Frequency Range: 500kHz to 200MHz
Output (Envelope) Frequency Range: 30Hz to 5kHz
Probe Input Capacitance (Approximate): 2.25pF
Probe Input Resistance (Approximate):

AT 500 kHz	25 KILOHMS
1 MHz	23 KILOHMS
5 MHz	21 KILOHMS
10 MHz	18 KILOHMS
50 MHz	10 KILOHMS
100 MHz	5.0 KILOHMS
150 MHz	4.5 KILOHMS
200 MHz	2.5 KILOHMS

MAXIMUM INPUT:
AC VOLTAGE .... 20 RMS VOLTS
28 PEAK VOLTS

Fig. 3-11. A standard demodulator-probe circuit, with response characteristics.

Shunt-Detector Probe With Two-Section RC Filter

A standard demodulator probe circuit, with response characteristics, is shown in Fig. 3-11. This is a shunt-detector configuration with a two-section *RC* filter; the coaxial cable provides the capacitance for the second section. Note that the 220-kilohm resistor serves as a filter component and also isolates the coaxial cable from the high-frequency circuit, thereby avoiding development of standing waves. Since the envelope frequency capability of the probe extends only to 5 kHz, a horizontal-sync pulse cannot be reproduced, although a vertical-sync pulse is reproduced when the demodulator probe is applied in a tv if circuit. A vertical-sync pulse, however, is distorted to the extent that its serrations and equalizing pulses are "wiped out." Thus the reproduced vertical-sync pulse has a super-

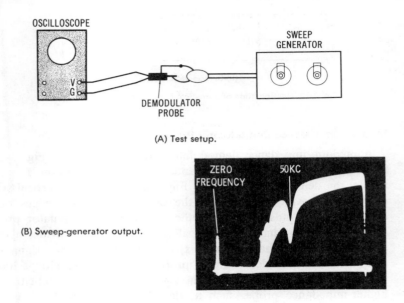

(A) Test setup.

(B) Sweep-generator output.

Fig. 3-12. Demodulator probe is used to check "flatness" of sweep-generator output.

ficial similarity to a horizontal-sync pulse in this situation. The demodulator probe has an input capacitance of approximately 2.25 pF and an input resistance of about 10 kilohms at 50 MHz. Accordingly, tv if circuits are substantially loaded and somewhat detuned by probe application. Therefore, the demodulator probe serves as a signal-tracing device but is not a reliable indicator of the signal-voltage level. However, a demodulator probe is an accurate *relative-signal-level* indicator when used in low-impedance circuits. For example, a demodulator probe is a reliable indicator to check the uniformity of output from a sweep generator (Fig. 3-12).

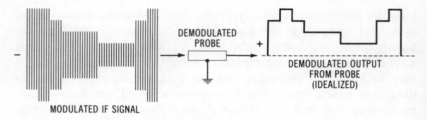

(A) Demodulating the if signal.

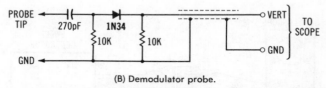

(B) Demodulator probe.

Fig. 3-13. Configuration of a medium-impedance demodulator probe.

Medium-Impedance Demodulator Probe

A medium-impedance demodulator probe, depicted in Fig. 3-13, is sometimes used to avoid substantial waveform distortion. For example, the demodulator probe in Fig. 3-11 reproduces vertical-sync pulses in the video signal, but the probe practically "wipes out" horizontal-sync pulses. On the other hand, the demodulator probe in Fig. 3-13 reproduces the vertical-sync pulses and provides a reasonable replica of the horizontal-sync pulses in the video signal. A medium-impedance demodulator probe imposes more circuit loading, however, and it attenuates the signal to a greater extent than a higher-impedance probe, such as depicted in Fig. 3-11. Thus the probe shown in Fig. 3-13 is a compromise design between circuit loading and waveform reproduction. Both types of demodulator probes are sometimes called *traveling detectors,* because they can be used to trace a signal stage by stage through an if amplifier section. As a practical note, signal tracing at the input (and occasionally at the output) of the first if stage may be impossible unless an oscilloscope with high sensitivity is used, such as 10 mV/cm.

Low-Impedance Demodulator Probe

Troubleshooting with the oscilloscope sometimes requires the use of a low-impedance demodulator probe. For example, when the frequency response of a single if stage is to be checked, a low-impedance demodulator probe is required, as shown in Fig. 3-14. A sweep signal is applied at the input of the stage to be checked, and a low-impedance demodulator probe is applied in the collector circuit of the following transistor. Note that a low-impedance demodulator probe can be easily devised by connecting a 300-ohm resistor across the input terminals of a conventional demodulator probe. The 300-ohm resistor "swamps" the frequency response of L3 so that the displayed curve essentially represents the response of the L2 stage. Note that the low output impedance of the sweep generator "swamps" the frequency response of L1.

All demodulator probes have a short ground-return lead in addition to the probe tip on the housing (case). It is important to *always connect this short ground lead from the probe housing to a chassis-ground point near the signal-takeoff point of the probe tip.* The reason for this requirement is that if frequencies are in the 45-MHz range, and a long ground lead to the oscilloscope will develop standing waves and resulting pattern distortion. Therefore the ground-return circuit must be kept short, and the high-frequency ground connection must be made via the black lead from the probe housing. It is also good practice to observe the chassis-ground point for the stage under test, and to ground the probe at this point. For example, if the probe tip is applied at the collector of a transistor, it is advisable to connect the black lead from the probe housing to the chassis-

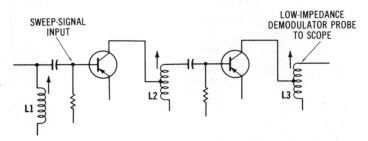

(A) Application in checking frequency response of single if stage.

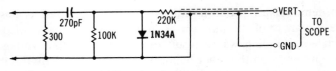

(B) Typical probe arrangement.

Fig. 3-14. Low-impedance demodulator probe.

ground point for the emitter of the same transistor. This technique avoids the hazard of interfering high-frequency ground currents.

Double-Ended (Push-Pull) Demodulator Probe

A double-ended demodulator probe, as depicted in Fig. 3-15, is preferred for checking the termination of a transmission line. (If a coaxial cable termination is to be checked, a conventional single-

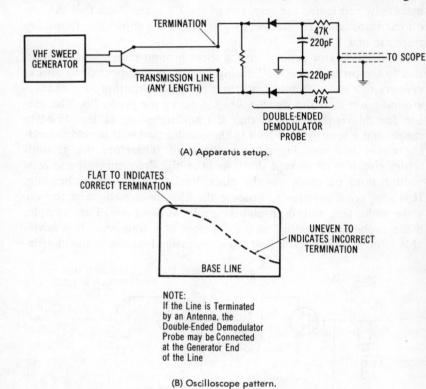

(A) Apparatus setup.

(B) Oscilloscope pattern.

Fig. 3-15. Configuration of a double-ended demodulator probe, and application for checking of lead-in termination.

ended demodulator probe is utilized.) This sweep test is based on the development of standing waves in case that the line is incorrectly terminated. Standing waves result in change of output voltage over the swept band. On the other hand, if the line is correctly terminated, no standing waves occur, and the oscilloscope pattern is the same as if the demodulator probe were applied at the output terminals of the sweep generator. The same method can be used to check whether a transmission line is correctly matched by an antenna, and whether matching stubs and similar devices are in proper adjustment.

RESISTIVE "ISOLATING" PROBE

A resistive "isolating" probe is a simple arrangement consisting of a resistor connected in series with the coaxial input cable to the oscilloscope, as shown in Fig. 3-16. Although called an "isolating probe," it is technically a low-pass RC filter section. It can be compared to an integrating circuit. This type of probe is used only in sweep-alignment procedures. Low-pass filtering action serves to sharpen the beat-marker (pip) indication on a response curve; it also serves to minimize noise (fuzz) on the curve when one is checking the response of low-level circuits. In most situations a 50-kilohm resistor is suitable; if the resistor is too large, the marker indication on the side of a response curve will be displaced (due to time delay). On the other hand, if the resistor is too small, the beat-marker indication will be broader than necessary (high beat frequencies are passed).

HIGH-VOLTAGE CAPACITANCE-DIVIDER PROBE

Although used primarily in laboratories, high-voltage capacitance-divider probes also find occasional application in tv troubleshooting procedures. High peak-to-peak voltages are encountered in the horizontal-sweep section of a tv receiver. These voltages will arc through a low-capacitance probe, damaging both probe and scope. A special probe therefore is required to test these high ac voltages. A typical circuit is shown in Fig. 3-17; this is a capacitance-divider arrangement. When two capacitors are connected in series, an applied ac voltage drops across the capacitors in inverse proportion to their capacitance values. Thus if one capacitor has 99 times the capacitance of the other, 0.01 of the applied voltage is dropped across the larger capacitor. In turn, the smaller capacitor requires a high voltage rating. The standard attenuation factor of a high-voltage capacitor-divider probe is 100 to 1, and is set by a trimmer capacitor. This 100-to-1 attenuation factor is used to tie the probe attenuation in with the step attenuator of the oscilloscope. The probe attenuates horizontal-sweep circuit waveforms to 0.01 of their source-voltage value, thus protecting the oscilloscope against damage.

This high-voltage capacitance-divider probe is uncompensated and is therefore useful only at horizontal-deflection frequencies. Vertical-frequency waveforms would therefore be distorted. This, however, is not a drawback, because the attenuation factor of the probe restricts its application to the horizontal-sweep circuitry. The reason why the 100-to-1 probe is unsuitable for vertical-section tests is shown in Fig. 3-18A. Observe that the probe capacitors do not stand alone but work into the vertical-input impedance (R_{in} and C_{in}) of

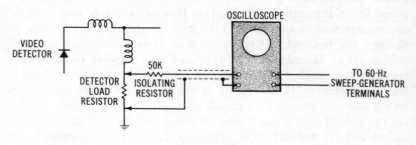

(A) Test connections.

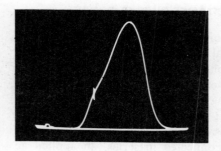

(B) If response curve with beat marker.

Fig. 3-16. Application of resistive "isolating"

the oscilloscope (Fig. 3-18B). Shunt resistance R_{in} can be neglected at the horizontal-deflection frequency (Fig. 3-18C), because this resistance value is very high compared with the low reactance of the input capacitance. On the other hand, at vertical-deflection frequency (Fig. 3-18D), shunt resistance R_{in} has a value on the same order as the reactance of the input capacitance. The capacitance-divider probe now acts as a differentiator of the vertical-frequency waveform, and the display is badly distorted.

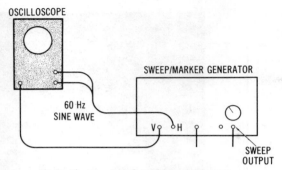

(C) Oscilloscope is deflected by 60-Hz sine wave.

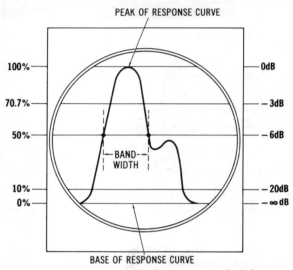

(D) Response curve amplitude in percent in decibels.

probe in sweep-alignment procedures.

Stray-Field Interference

Some older-model oscilloscopes have exposed VERTICAL INPUT binding-post terminals, although a shielded cable is used with the associated low-capacitance probe or demodulator probe. These ex-

Fig. 3-17. Typical high-voltage capacitor-divider probe.

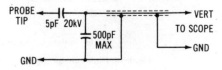

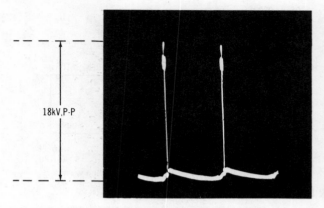

(A) Distortion of pulse at vertical-deflection frequency.

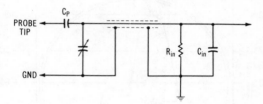

(B) Configuration when connected to vertical input of oscilloscope.

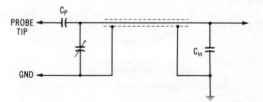

(C) Equivalent circuit of (B) at high frequencies.

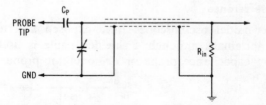

(D) Equivalent circuit of (B) at low frequencies.

Fig. 3-18. High-voltage capacitor-divider probe and its load circuit.

posed binding posts can be responsible for stray-field interference in patterns; the interference may have the form of 60-Hz hum or of flyback pulses. The reason for stray-field pickup is that the exposed VERTICAL INPUT terminal has high impedance, even though a low-impedance receiver circuit may be under test. For this reason virtually all modern oscilloscopes are provided with coaxial connectors for the vertical input. Note in passing that 60-Hz hum interference in displayed patterns is not always the result of stray-field pickup. For example, the ground-return lead to a low-capacitance probe or a demodulator probe could be defective. Beginners are sometimes puzzled by the strong interference produced from nearly fluorescent lamps.

Inconsistent Low-Capacitance Probe Response

It is occasionally observed that although a low-capacitance probe is adjusted for proper response on one setting of the VERTICAL step attenuator, its response is more or less incorrect on another setting. This difficulty results from improper adjustment of the compensating trimmers in the VERTICAL step attenuator. In this situation the compensating trimmer capacitors must first be adjusted correctly on the basis of a 15-kHz square-wave signal, using a direct probe. After proper square-wave response is provided by the VERTICAL step attenuator, the low-capacitance probe may then be adjusted for distortionless square-wave reproduction.

OVERVIEW OF PROBE APPLICATION IN TV CIRCUITRY

In Fig. 3-19 a simplified block diagram of a black-and-white tv receiver is shown. Various stages and test points for application of an oscilloscope probe are indicated. The letters at each test point denote the correct type of probe to use, and also indicate the approximate deflection rate for the oscilloscope: R, a direct probe; D, a demodulator probe. The letter H indicates a 7875- or a 15,750-Hz deflection rate, V indicates a 20-, 30-, or 60-Hz deflection rate, and A indicates an audio deflection rate, such as 200 Hz or 500 Hz. For clarity, all amplifier stages of a given type are included within one block in the diagram of Fig. 3-19. However, tests may be made at input terminals or output terminals of individual amplifier stages, by using the indicated probe and deflection rate. Note that at any point up to the picture detector, the signal voltages will be comparatively small (and may be very small) so that considerable vertical

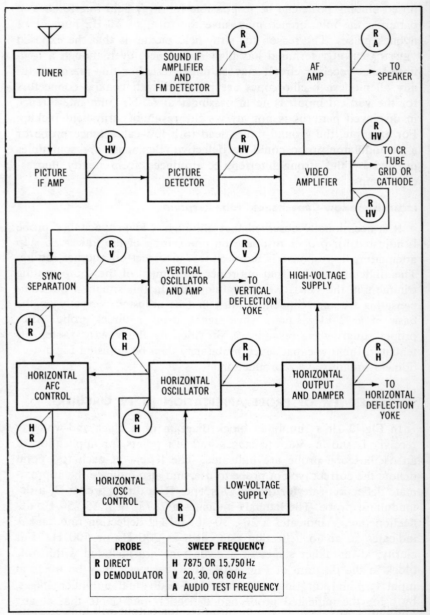

Courtesy Heath Co.

Fig. 3-19. Simplified block diagram of a tv receiver, showing test points.

gain is required in the oscilloscope. On the other hand, within the sync and deflection circuits, signal voltages are comparatively high, so much less vertical gain is required in the scope. When you are troubleshooting with the oscilloscope, remember that a circuit or device malfunction often causes normal signal amplitudes to be greatly weakened or perhaps reduced to zero.

Although many useful tests can be made with the receiver energized by a station signal from an antenna, some waveform checks are facilitated by the use of various generators. A test-pattern generator provides a steady signal that can be increased or decreased in level as desired. Checks of tuned-circuit frequency response require the use of a sweep-and-marker generator. A color-bar generator is essential for troubleshooting chroma circuitry. Hi-fi stereo troubleshooting requires an accurate audio generator and a stereo-multiplex generator. Various oscilloscope tests in CB transceivers require the availability of a lab-type signal generator. Square-wave and pulse generators also facilitate troubleshooting procedures with the oscilloscope.

VERTICAL INTERVAL TEST SIGNAL

A useful and highly informative check of picture-channel frequency (and phase) response is provided by the vertical interval test signal (VITS). With reference to Fig. 3-20, tv station signals are often supplemented by the VITS signal, which is transmitted on two lines of the vertical-sync interval. This signal is a multiburst waveform with frequency steps of 0.5, 1.5, 2.0, 3.0, 3.6, and 4.2 MHz. The multiburst is followed by a sine^2 pulse, and by a window pulse (white flag). On the next scanning line the multiburst signal is repeated, followed by an 11-step staircase waveform; each step of the staircase signal has a superimposed color burst. This VITS signal can be picked out of the composite video signal by a good-quality triggered-sweep oscilloscope. Distortions undergone by the VITS waveform at the output of the picture detector indicate deficiencies in rf/if frequency and phase response. Distortions in the waveform at the output of the video amplifier indicate additional deficiencies in the video-amplifier frequency and phase characteristic.

The multiburst pattern is of primary interest to the tv troubleshooter. Thus, in case that the rf/if frequency response is poor, the multiburst waveform at the picture-detector output will appear more or less as exemplified in Fig. 3-20C. That is, the higher-frequency

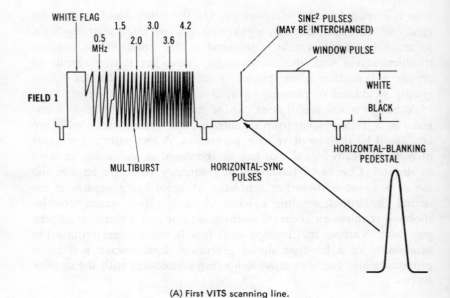

(A) First VITS scanning line.

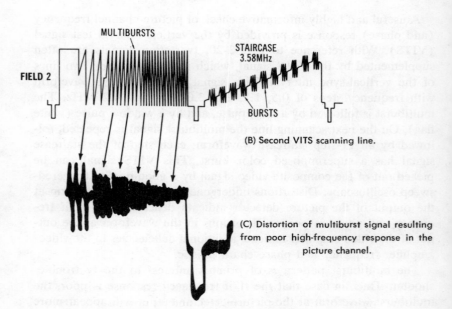

(B) Second VITS scanning line.

(C) Distortion of multiburst signal resulting from poor high-frequency response in the picture channel.

Fig. 3-20. Vertical interval test signal.

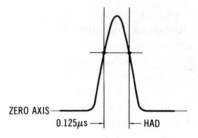

ZERO AXIS

0.125μs — | |— HAD

(A) Transmitted waveform with half-amplitude duration (had) of 0.125 μs.

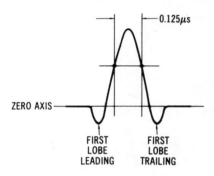

|— 0.125μs

ZERO AXIS

FIRST LOBE LEADING

FIRST LOBE TRAILING

(B) Normal reproduction of pulse by receiver circuits.

Fig. 3-21. Sine² pulse waveforms.

bars will be substantially attenuated. On the other hand, if the rf/if bandwidth is approximately 4 MHz, all bars with the exception of the 4.2-MHz bar will be reproduced at full amplitude. The sine² pulse is informative, because it has the approximate shape of a single picture element; therefore, if the sine² pulse is reproduced without objectionable distortion, the elements in the camera signal will also be properly reproduced. The window signal (white flag) is 18 μs in duration, and will show ringing distortion in the event of an improper phase characteristic in the picture channel. Finally, the staircase signal is useful to indicate linearity (or lack of linearity) in the picture channel. The reproduced staircase waveform should not show curvature either at the bottom or at the top of the ramp. Fig. 3-21 shows the sine² pulse waveform as transmitted, and its normal reproduction through the tv receiver circuits.

SPECIAL TYPES OF OSCILLOSCOPE PROBES

Although used chiefly in tv labs, special types of oscilloscope probes are available to extend the range of applications for both conventional and for specialized oscilloscopes. Thus a 100-to-1 low-capacitance probe may be utilized; current probes permit display of current waveforms on the scope screen; high-voltage capacitance-divider probes provide display of waveforms in high-voltage circuitry up to 40 kV; differential probes convert a single-ended oscilloscope into a double-ended vertical-input oscilloscope; logarithmic probes convert a waveform from "rectangular coordinates" to "semilogarithmic coordinates"; and various digital-logic probes provide greatly increased test capabilities in computer circuitry. Some types of logic probes include digital-signal processing facilities for recognition of unique digital events.

CHAPTER 4

Signal Tracing in RF, IF, and Video Amplifiers

Signal tracing with the oscilloscope is the procedure by which the progress, amplitude, and characteristics of an applied signal voltage are checked, stage by stage, through the signal channels of a receiver (a television receiver in this instance). The signal channels comprise an rf amplifier, mixer, video-if amplifier, video amplifier, sound-if amplifier, and audio amplifier in a black-and-white tv receiver (Fig. 4-1). In a color tv receiver the signal channels also include a band-pass amplifier and various chroma-signal processing sections. Note in Fig. 4-1 that an rf input voltage of 150 μV will normally produce a video-amplifier output voltage of 45 V when the receiver is operating at maximum available gain. This is a total signal-voltage gain of 300,000. In typical situations, however, the receiver does not operate at maximum available gain; the rf input voltage is typically 1500 μV, and the total signal-voltage gain 30,000. The if amplifier and the rf amplifier operate at reduced gain as the rf input voltage increases, due to agc action.

TROUBLESHOOTING RF AMPLIFIERS

When a television receiver has the trouble symptom of "no picture and no sound" and the snow level is high, signal tracing with the oscilloscope starts logically at the tuner—most of the snow is intro-

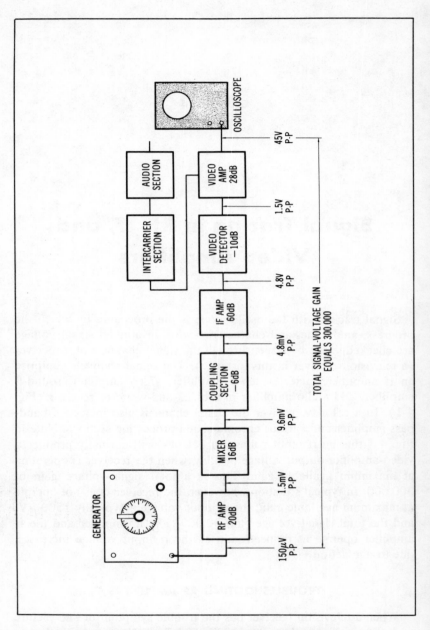

Fig. 4-1. Block diagram of typical black-and-white tv receiver operating at maximum available gain.

duced by the mixer stage. Many tv receivers provide a "looker point" for the mixer input circuit (see Fig. 4-2). An oscilloscope may be applied at the looker-point terminal on the tuner to check the rf amplifier for normal operation. A vhf sweep generator signal is applied to the antenna-input terminals of the tuner, and a dc bias override voltage of suitable value is connected to the agc line of the tuner. The receiver service data will often note the normal agc bias-voltage value. Note that the mixer input circuit operates as a detector; in turn, the rf frequency-response curve is displayed on the oscilloscope screen. This test reveals several important facts:

1. It shows whether a vhf signal is passing through the rf amplifier.
2. It shows whether the rf amplifier has approximately normal gain.

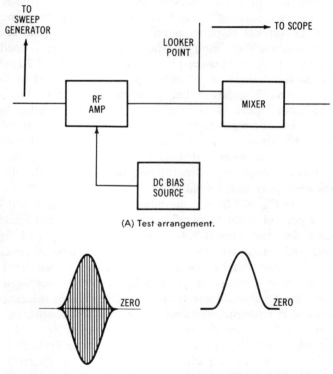

(A) Test arrangement.

(B) High- and low-frequency signals at "looker point."

Fig. 4-2. Oscilloscope may be applied at "looker point" in the mixer input circuit.

3. It shows whether the rf amplifier has normal bandwidth, or whether the response curve may be seriously distorted.

In case that the rf amplifier checks out normally but no picture is displayed although there is a high snow level, the local oscillator in the front end is probably "dead," or operating considerably off frequency. For a quick cross-check, transfer the oscilloscope to the output of the picture detector. Absence of an overall response curve at the output of the picture detector, with a high noise level, throws strong suspicion on the local oscillator. On the other hand, there is a chance that an overall response curve might be observed at the picture-detector output, with little or no observable noise. This test result indicates that a transistor in the video-amplifier section is defective and is generating high-level noise voltages. Such a transistor is usually found to have considerable collector-junction leakage.

It follows from previous discussion that if an oscilloscope is used with a resistive "isolating" probe, the effective high-frequency response of the vertical amplifier is reduced, but the vertical sensitivity will be practically the same as when the oscilloscope is used with a direct probe. If the oscilloscope has a rated sensitivity of 20 mV p-p per inch, an input signal of 60 mV p-p will be required to display a response curve 3 inches in height. With reference to Fig. 4-2 a sweep-generator signal of approximately 20,000 μV must be applied at the antenna-input terminals of the tuner. Note that this is a comparatively "strong" input signal; however, this signal level will not overload the rf stage and distort the response curve, provided that the tuner is operating normally and provided that the dc override bias voltage is suitably adjusted.

As shown in Fig. 4-2 both demodulated and undemodulated sweep signals are present at the "looker" point. The mixer input circuit operates as a detector; various heterodyne products (types of signals) are developed. Harmonic frequencies and sum-and-difference frequencies are produced by the mixer; the demodulated frequency-response curve is a difference product between the sweep-generator signal and the local-oscillator frequency. Reasonable tolerances on the shapes of rf frequency-response curves are permissible, as exemplified in Fig. 4-3. Both the picture-carrier marker and the sound-carrier marker normally appear on top of the curve. Note that if the amplitude of the response curve on one or more channels were greatly subnormal, a tuner defect would be indicated. Or, if a response curve on one or more channels should happen to have an

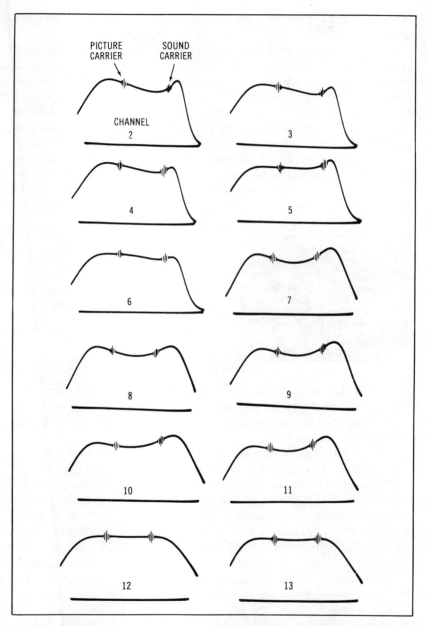

Fig. 4-3. Rf response curves showing normal tolerances on the 12 vhf channels for a good-quality receiver.

excessive "hill-and-valley" contour, it would be concluded that the tuner is faulty.

When a frequency-response curve as shown in Fig. 4-4 is observed at the "looker" point, it is indicated that the rf amplifier is regenerative (often due to an open decoupling or bypass capacitor). The response curve is highly peaked, is often "looped" at one or both ends, and is very unstable. If the operator brings his or her hand near the "looker" point, the curve changes shape. It will also be observed that the curve shape depends considerably upon the value of override bias voltage that is utilized. As the bias voltage is changed to increase the stage gain, the curve becomes more peaked, and at some critical bias value the rf amplifier breaks into uncontrolled oscillation. Oscillation blocks the sweep-signal flow, and only a horizontal line is dis-

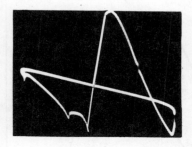

Fig. 4-4. Typical frequency-response curve produced by a regenerative rf tuner.

played on the oscilloscope screen. A tuner that is seriously regenerative on the low channels may operate normally on the high channels. Decoupling and bypass capacitors for a standard tuner are indicated in Fig. 4-5.

Subnormal or zero gain on one or more channels may often be tracked down to defective switch contacts. Once in a while a seemingly defective tuner has merely been seriously misaligned by a "do-it-yourselfer." If lightning strikes near an antenna that does not have a lightning arrester, coil and/or device damage can result in the rf-amplifier circuitry. Intermittent defects are more obscure and are often very difficult to locate; cold-soldered connections, worn switches, microscopic cracks in printed-circuit conductors, marginal breakdown in capacitors, or failing transistors are typical culprits. Some tv repair shops have suitably experienced technicians who can repair defective tuners. Most shops, however, prefer to send a defective tuner to a specialized repair depot, such as Precision Tuner Service, Inc., or Castle TV Tuner Service, Inc.

With reference to Fig. 4-2, note that no display will be obtained on the oscilloscope screen if the resistive "isolating" probe is transferred from the "looker" point to the mixer output terminal. The reason for this seeming lack of signal output is due to the fact that the mixer output circuit is tuned to pick out the if component from the heterodyne products. Thus, in most receivers the demodulated frequency-response curve is completely rejected—only if the if signal is present at the mixer output, to which the oscilloscope cannot directly respond. Of course, if the resistive "isolating" probe is transferred ahead to the picture-detector output, the overall rf/if response curve will then be displayed on the oscilloscope screen.

SIGNAL TRACING IN THE IF SECTION

As noted previously, a service-type oscilloscope cannot respond directly to a tv if signal because its vertical-amplifier frequency response is inadequate. Therefore a demodulator probe must be used to trace the signal through the if strip. With reference to Fig. 4-6 the lowest signal level occurs at point A, and the highest signal level occurs at point D. Note that a receiver with three if stages has a typical rf/if gain of 100,000 when a very weak signal, such as 20 μV, is applied to the antenna-input terminals. On the other hand, when a strong signal, say 20,000 μV, is applied to the antenna-input terminals, agc action reduces the rf/if gain to 100. Thus the gain of the rf/if system normally varies over a range of 1000 to 1 (60 dB) in accordance with the input signal level.

Although a tv station signal can be used in if signal-tracing procedures, it is advantageous to use a steady and controllable test signal, as from a pattern generator. Unless a high-level input signal is used, it is often difficult to determine whether or not signal voltage is present at point A in Fig. 4-6. In other words, conventional modulator probes impose a substantial insertion loss, and, unless the oscilloscope has quite high sensitivity, the signal level at point A must be abnormally high to produce a recognizable pattern on the crt screen. Of course, it is desirable in many situations to know whether or not a signal voltage is present at point A—this information serves to distinguish between faults in the base circuit and in the collector circuit of the first if stage. Note in passing that a conventional demodulator probe rejects the higher video-frequency components and passes only the low-frequency video components, as shown in Fig. 4-7. So, horizontal-sync pulses, equalizing pulses, and serrations are

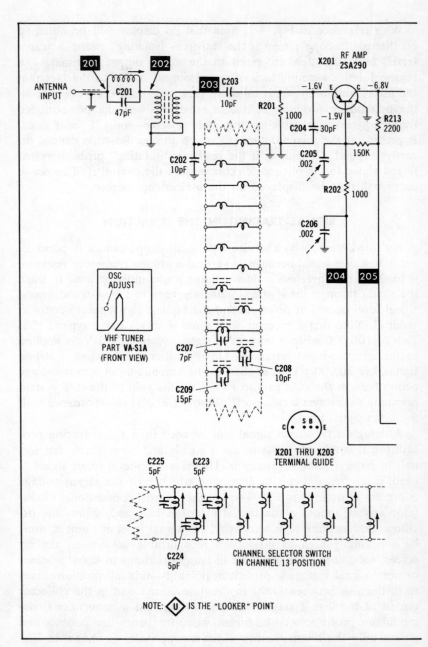

Fig. 4-5. Standard tuner configuration. Dotted

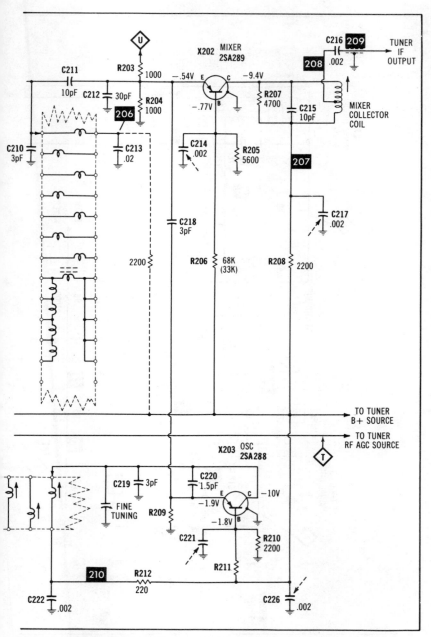

arrows point to decoupling and bypass capacitors.

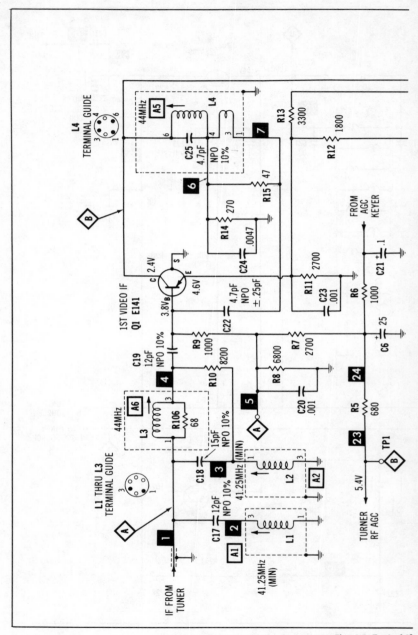

Fig. 4-6. Typical if

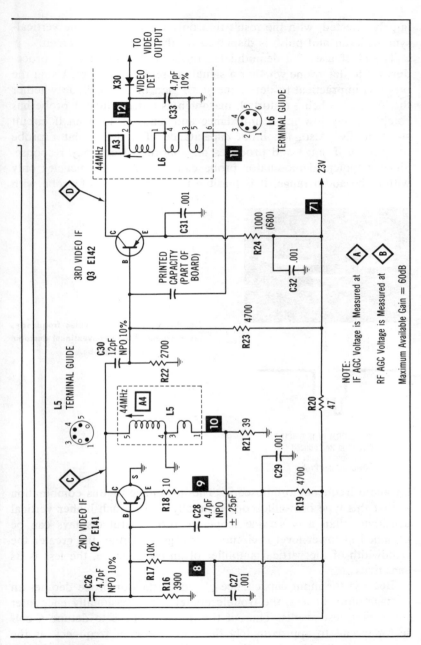

amplifier configuration.

largely rejected, with the result that only the outline of the vertical-sync pedestal and pulse is displayed on the oscilloscope screen.

The chief use of a demodulator probe in if signal-tracing procedures is to determine whether a signal is present or absent. Using the probe is impractical to determine stage gain or to conclude whether the if signal is being distorted, inasmuch as a demodulator probe has comparatively low input impedance and tends to load an if circuit substantially. Also, the input capacitance of a demodulator probe detunes an if stage and produces an abnormal frequency response. Since a typical demodulator probe can reproduce frequencies only within the audio range, it is feasible to use this type of probe with

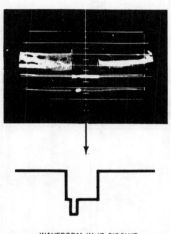

Fig. 4-7. Only the lower video frequencies are reproduced by a conventional demodulator probe.

WAVEFORM IN IF CIRCUIT
Waveform Reproduced
by Conventional
Demodulator Probe

an audio-frequency oscilloscope. The advantage of this combination is that this type of oscilloscope generally has a much higher vertical sensitivity than a tv service scope. In turn, useful displays can be obtained in lower-level if circuits. As a general rule, the greater the bandwidth of the vertical amplifier of an oscilloscope, the less is its sensitivity.

Because the input capacitance of a demodulator probe detunes an if stage more or less, the troubleshooter will occasionally encounter misleading test results. Thus, if the probe happens to detune a collector circuit to approximately the same resonant frequency as the base circuit, the if stage may "take off" and oscillate uncontrollably.

As a result, signal passage is blocked through the stage, and the troubleshooter concludes that the stage is "dead." However, if the probe is moved ahead to the following stage, a signal pattern is unexpectedly observed. Note that if oscillation is accompanied by a substantial dc output voltage from the picture detector. Next, another misleading test result occurs when a demodulator probe detunes an if circuit enough to make the stage regenerative, without breaking through into oscillation. In such a case the stage gain becomes abnormally high, and the stage frequency response is greatly reduced in bandwidth. Waveform distortion is difficult or impossible to discern in this situation, with the result that the troubleshooter is likely to conclude that the particular if stage has unusually high gain for some unexplained reason. It is helpful to note that a regenerative stage is highly unstable—if the operator brings his or her hand near the probe, the display on the oscilloscope screen will change considerably.

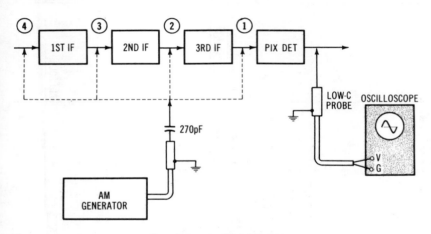

Fig. 4-8. Stage-gain test setup.

Follow-up Tests and Cross-Checks in IF Signal Tracing

To repeat an essential point, if signal-tracing procedures often serve to pinpoint the trouble area, but have limited capability in pinpointing defective components, faulty devices, or circuit misadjustments. Occasionally an open coupling capacitor can be localized —the if signal is present at the input end of the capacitor, but is absent at the output end. In most cases, however, initial clues must be investigated by follow-up tests and/or cross-checks. For example, suppose that a trouble symptom of "weak picture" is being tracked down. Signal-tracing if tests are often inconclusive and should be followed up by signal-injection tests, as shown in Fig. 4-8. An am signal generator is suitable. The generator is tuned to midband frequency (approximately 44 MHz, as in Fig. 4-9). With the oscillo-

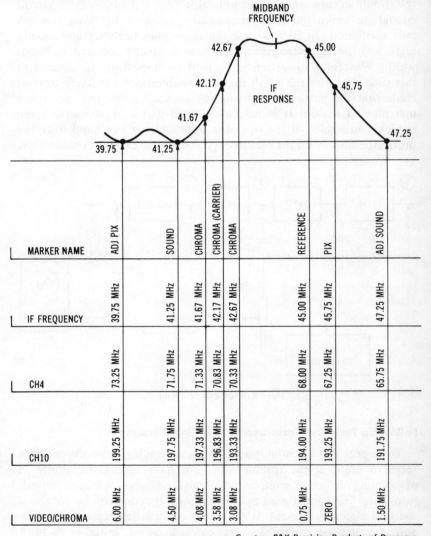

Courtesy B&K Precision Products of Dynascan

Fig. 4-9. Frequency relations along a standard if response curve.

scope applied at the output of the picture detector with a low-capacitance probe, the generator signal is applied successively at points 1, 2, 3, and 4. Overloading must be carefully avoided—in other words, pattern clipping is not permitted when true gain values are sought.

To start the stage-gain test depicted in Fig. 4-8, the if agc bias line is clamped at a moderate value (this value may be indicated in the receiver service data). Then, with the generator signal applied at point 1 and with at least 30-percent modulation, the signal level is adjusted to obtain a usable pattern on the oscilloscope screen. The pattern height should be fairly low, however, because the generator output cable will next be transferred to point 2, without any reduction in the generator output level. When this transfer is made, the operator will normally see a very substantial increase in pattern height—its ratio to the previous pattern height is a measure of the stage gain. For example, the ratio of pattern heights may be in the range from 5 to 10, depending on the value of agc override bias that is utilized. Note that in many receivers the last if stage is not agc-controlled, however; in such a case the last if stage always operates at maximum gain, regardless of the agc bias value.

Next, the gain of the second if stage is checked. To avoid overloading, the output level from the signal generator (connected at point 2) is reduced to obtain a low-amplitude but usable pattern on the oscilloscope screen. Then the generator output cable is transferred to point 3, and the change in pattern height is observed; this change indicates the gain of the second if stage. Finally, the gain of the first if stage is determined in similar manner. When the three stage-gain values are "sized up" it may be observed that one of the values is very low, compared with the other two values. In consequence, the troubleshooter will conclude that the "weak picture" symptom is being caused by a defect in the stage that exhibits subnormal gain. Then dc-voltage measurements are usually made to assist in pinpointing the defective component or device.

Poor Picture Quality

The most common type of "poor picture quality" is lack of detail (sharpness), and this symptom casts suspicion on if alignment—most of the gain and selectivity of the signal channel is provided by the if amplifier. Fig. 4-9 shows that the standard if bandwidth from the 42.17-MHz to the 45.75-MHz points on the response curve is equal to 3.58 MHz. Subnormal bandwidth causes attenuation or re-

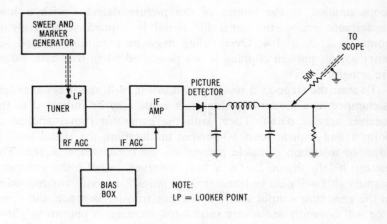

(A) Connections.

NOTE:
LP = LOOKER POINT

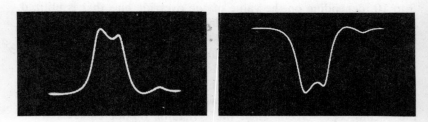

(B) Two aspects of the oscilloscope pattern.

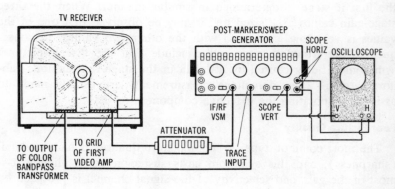

Courtesy Heath Co.

(C) Equipment arrangement for post-marker generator.

Fig. 4-10. Typical if sweep-alignment test setup.

jection of the higher video frequencies, thereby impairing reproduction of detail in the picture. To check the if bandwidth, connect the oscilloscope through a 50-kilohm resistor to the output of the picture detector, and apply an if sweep-and-marker test signal to the looker point on the tuner, as shown in Fig. 4-10A. Both the rf and if agc lines should be clamped at normal bias levels. Note that whether a positive-going or a negative-going if response curve is displayed depends on the polarity of the picture-detector diode.

Fig. 4-11. Black and white shades are reversed in a negative picture.

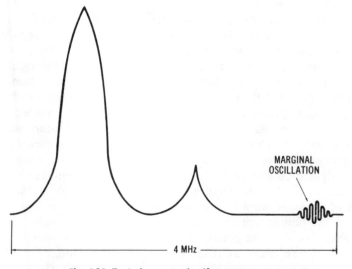

MARGINAL
OSCILLATION

4 MHz

Fig. 4-12. Typical regenerative if response curve.

The if bandwidth is checked by placing a marker half-way up the left side of the curve, and noting the marker frequency. Then the marker is moved to the half-way point on the right side of the curve, and its frequency is noted. The difference between these two marker frequencies is the bandwidth of the if response curve. If the bandwidth is substantially incorrect, the troubleshooter will proceed to realign the if amplifier according to the instructions provided in the receiver service data. As a practical note, realignment should not be undertaken until after all troubleshooting has been completed. This is because misalignment often occurs as the result of a component defect in the if circuitry; after the defect has been corrected, it is likely that the alignment curve will also be normal.

Observe in Fig. 4-10B that the if response curve may be either positive going or negative going, depending on the polarity of the picture-detector diode. Accordingly, if the detector diode is defective and a polarity error is made in replacement, a *negative picture* symptom will result, as shown in Fig. 4-11. Normally white lines and areas appear as black, and normally black lines and areas appear as white in a negative picture. Note that a negative picture does not lock tightly in sync and that a display as illustrated in Fig. 4-11 may be obtained only when the HORIZONTAL HOLD and VERTICAL HOLD controls are critically adjusted. Although there are several possible causes of negative picture reproduction, reversal of the picture-detector diode polarity is the most likely cause.

Poor picture quality is sometimes caused by regeneration in the if amplifier. A regenerative overall if response curve is depicted in Fig. 4-12. Note that the responses are narrow and sharply peaked; in this example, marginal oscillation is occurring near the right-hand end of the pattern. Picture symptoms caused by if regeneration range from lack of detail through lack of solid areas accompanied by picture ringing ("circuit ghosts"). Another aspect of if regeneration is self-generated interference similar to CB or other rf interference; it is not a stable trouble symptom, but changes in its appearance from one channel to another. Sometimes a regenerative if strip will break into sustained oscillation (particularly during weak-signal reception), with the result that the picture-tube screen goes blank, without any trace of snow. When a test setup is utilized as shown in Fig. 4-10A, the troubleshooter will observe that the oscilloscope pattern is very unstable, and changes greatly whenever a test lead is moved. As in the case of a regenerative rf tuner, if regeneration and/or oscillation is most likely to be caused by an open decoupling or bypass capacitor.

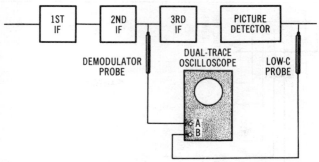

(A) A dual-trace scope monitoring the output of the 2nd if stage and the output of the picture detector.

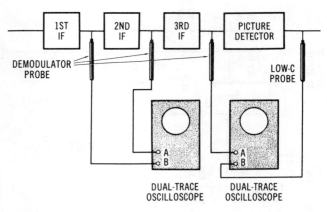

(B) Two dual-trace scopes monitoring four signal points in the if section.

Fig. 4-13. Dual-trace oscilloscopes can monitor two or more signal points simultaneously.

However, this trouble symptom can also occur in the event that a do-it-yourselfer has tampered with the if alignment adjustments. If the base circuit and collector circuit of an if transistor are resonated at the same frequency, regeneration and/or oscillation is very likely to occur.

Signal-Tracing IF Intermittents

The most difficult tough-dog problems are often encountered in intermittent situations, where the picture is normal for a period and then suddenly or gradually disappears. After a longer or shorter interval the picture will just as inexplicably reappear. In other types of

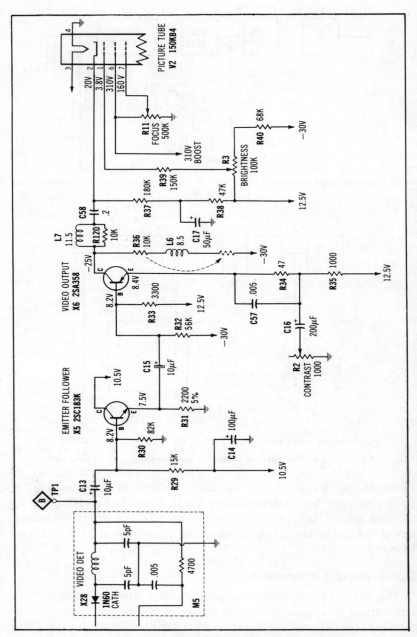

Fig. 4-14. Typical transistor video amplifier.

intermittent trouble symptoms the picture will be normal for a while and then become distorted—it may become blurred, or interference may be displayed. Then, after some time has passed, the picture will again be reproduced normally. Preliminary troubleshooting procedure in such cases is concerned with the question: Where is the intermittent fault located? As noted previously, intermittent conditions can occur in the tuner—and they can occur as well in the if amplifier. When the if amplifier falls under suspicion the most efficient troubleshooting approach is *intermittent monitoring*. This is a specialized type of signal tracing, in which the if signal is continually observed at more than one point in the if strip.

As depicted in Fig. 4-13, if a pair of dual-trace oscilloscopes are available, they can be used to monitor four circuit points simultaneously. When an intermittent occurs the trouble can be localized im-

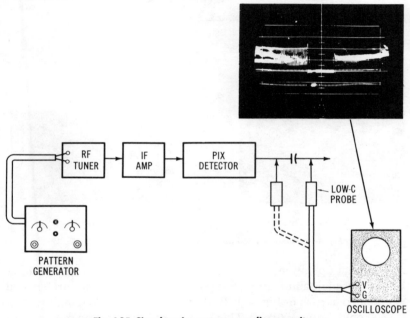

Fig. 4-15. Signal tracing across a coupling capacitor.

mediately to a particular stage. Note that monitoring is much more effective than conventional signal tracing in many intermittent situations, because monitoring procedure does not introduce transients or change the circuit operating conditions in any way. That is, the mere application of a test probe may "trigger" a failed intermittent circuit

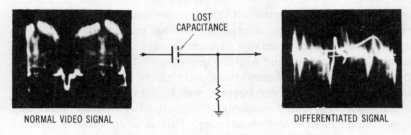

LOST CAPACITANCE

NORMAL VIDEO SIGNAL

DIFFERENTIATED SIGNAL

(A) Differentiation of video signal by defective coupling capacitor.

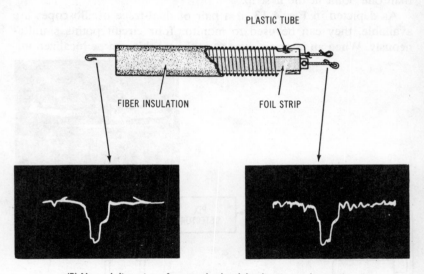

PLASTIC TUBE

FIBER INSULATION

FOIL STRIP

(B) Normal distortion of sync pulse by delay line in a color receiver.

Fig. 4-16. Abnormal and normal distortions of the video signal.

into normal operation, so that the oscilloscope operator gets no indication of signal failure at the test point. When four probes are applied simultaneously, as in Fig. 4-13B, there is a potential loading problem that should be minimized. This is accomplished by operating both oscilloscopes at maximum vertical gain, and capacitively coupling the probes to the higher-level signal points.

A quick and easy method of capacitively coupling a probe to a signal takeoff point is to place one or more layers of masking tape around the associated terminal. Then the probe is clipped over the tape, instead of making direct contact with the terminal. The vertical

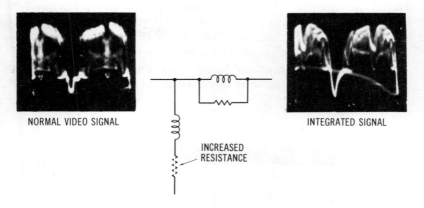

NORMAL VIDEO SIGNAL INCREASED RESISTANCE INTEGRATED SIGNAL

Fig. 4-17. Integration of the video signal by a load resistor that has increased considerably in value.

attenuation is therefore effected at the signal takeoff point instead of inside the oscilloscope. Although this method aggravates waveform distortion and may introduce noticeable hum into the pattern, it serves the particular purpose, because the troubleshooter is concerned only with indication of the signal-stopping area when the intermittent occurs. Some intermittents do not occur for long periods. In this case the troubleshooter can turn his or her attention to other activities, and check the monitor oscilloscope screens after some time has passed to see if a change has occurred. Sometimes the onset of an intermittent can be speeded up by mechanical jarring, by turning the power switch on and off, by blowing hot air on suspected components, or by spraying them with a coolant.

SIGNAL TRACING IN THE VIDEO AMPLIFIER

A low-capacitance probe is used when signal tracing in the video-amplifier section. A basic circuit for a video amplifier is shown in Fig. 4-14. This is an example of an ac-coupled amplifier; dc-coupled amplifiers are also in wide use. Coupling capacitors may become defective in ac-coupled amplifiers and cause various trouble symptoms. A coupling capacitor may become open and pass no video signal. Signal tracing at the input end and at the output end of the capacitor will quickly show whether the signal is passing through the capacitor, as illustrated in Fig. 4-15. If the video signal disappears when the low-capacitance probe is transferred from the input end to the output end of the capacitor, it means that the capacitor is open.

Sometimes a coupling capacitor loses a large portion of its capacitance but does not become completely open; this is typical of electrolytic coupling capacitors. In such a case the waveform at the input end of the capacitor is displayed normally, but the waveform at the output end of the capacitor is displayed at considerably reduced am-

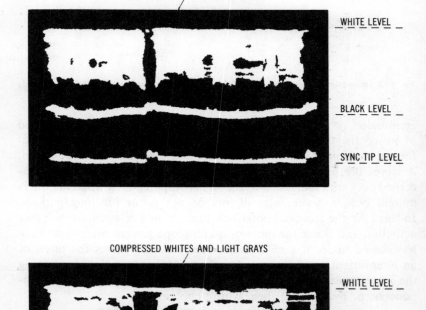

Fig. 4-18. Examples of white compression; sync pulse occupies more than 25 percent of waveform amplitude.

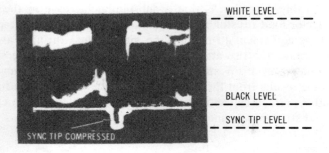

WHITE LEVEL

BLACK LEVEL

SYNC TIP LEVEL

SYNC TIP COMPRESSED

Fig. 4-19. Example of sync compression; sync pulse occupies less than 25 percent of waveform amplitude.

plitude and with substantial distortion. This distortion consists of waveform differentiation, as exemplified in Fig. 4-16A. Observe how the horizontal-sync pulse is changed into positive-going and negative-going spikes; the camera signal is also greatly distorted. Next, if a good capacitor is bridged across the defective coupling capacitor, the output waveform will be restored to normal. Observe in Fig. 4-14 that when a test capacitor is bridged across C13 or C15, it is essential to observe proper terminal polarity.

Next, suppose that a video-amplifier load resistor, such as R36 in Fig. 4-14, becomes defective and increases considerably in value. In this situation, signal tracing from base to collector of the output transistor will show high gain (higher than normal), but the output waveform will be distorted (integrated), as shown in Fig. 4-17. The

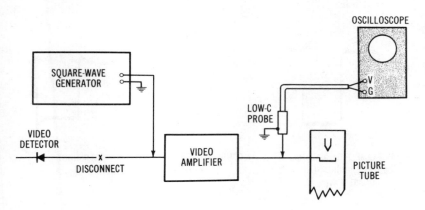

Fig. 4-20. Test setup.

reason for this distortion is that a substantial increase in the value of a collector load resistor results in loss of high-frequency response. This loss is equivalent to passing the video signal through an integrating circuit—the rise and fall times of the sync pulse are slowed down, and the sync tip appears "feathered." Note that integration in this case results from the combination of the collector load resistance and the circuit capacitances consisting of device junction capacitances and stray capacitances.

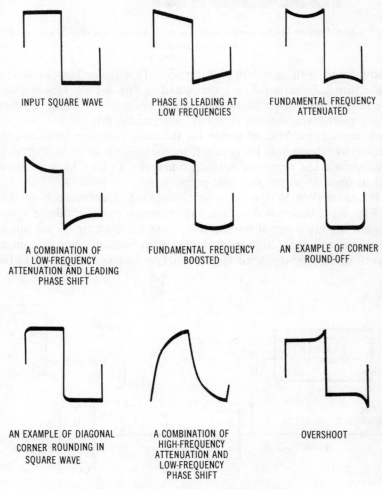

INPUT SQUARE WAVE

PHASE IS LEADING AT LOW FREQUENCIES

FUNDAMENTAL FREQUENCY ATTENUATED

A COMBINATION OF LOW-FREQUENCY ATTENUATION AND LEADING PHASE SHIFT

FUNDAMENTAL FREQUENCY BOOSTED

AN EXAMPLE OF CORNER ROUND-OFF

AN EXAMPLE OF DIAGONAL CORNER ROUNDING IN SQUARE WAVE

A COMBINATION OF HIGH-FREQUENCY ATTENUATION AND LOW-FREQUENCY PHASE SHIFT

OVERSHOOT

Fig. 4-21. Some basic square-wave distortions.

Waveform Compression and Clipping

A video amplifier normally operates in a linear manner: The output voltage is proportional to the input voltage at any point in the operating range. Sometimes, however, a defect in the video amplifier causes it to develop amplitude nonlinearity. An example is shown in Fig. 4-18. A test-pattern waveform is being displayed. This waveform has peak whites and represents the complete operating range of the video amplifier. Therefore the sync tip would normally occupy 25 percent of the waveform amplitude; as shown in the illustration, however, the sync tip occupies approximately 38 percent of the dis-

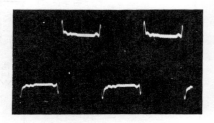

Fig. 4-22. Normal 100-kHz square-wave response of a high-performance video amplifier.

played waveform amplitude. This disproportion indicates that the white excursion of the waveform is being compressed and/or clipped. The result of this form of distortion is muddy, filled-up whites in the reproduced picture. The waveform distortion is called *white compression* (or *clipping*). Incorrect bias voltage on a video-amplifier transistor often causes this trouble symptom. Another typical possibility causing white compression is a marginal transistor with collector-junction leakage.

Another example of video-amplifier nonlinearity is shown in Fig. 4-19. Again a test-pattern waveform is being displayed; it has peak whites and represents the complete operating range of the video amplifier. Although the sync tip would normally occupy 25 percent of the waveform amplitude, it occupies only 16 percent, approximately, in this example. This disproportion indicates that the sync-tip excursion of the waveform is being compressed and/or clipped. The gray scale is distorted in the picture; dark grays are reproduced as light grays, and light grays are reproduced as white areas. The image tends to appear "washed out." This waveform distortion is called *sync compression* (or *clipping*). As in the previous trouble situation, incorrect bias voltage on a video-amplifier transistor is a common culprit. There is also the possibility of a failing transistor.

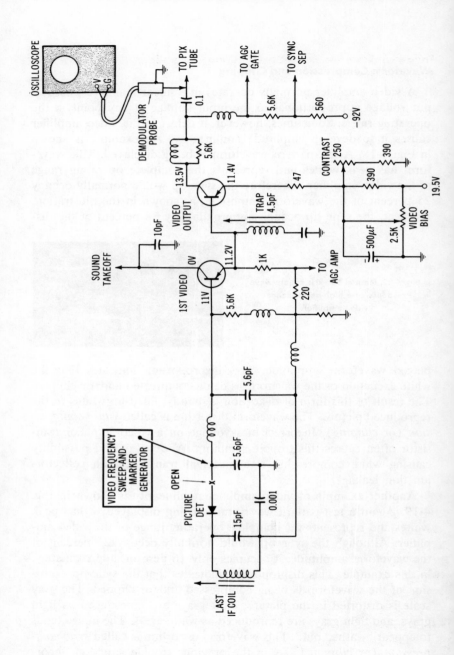

(A) Test connections.

Fig. 4-23. Sweep-frequency

Follow-up Tests and Cross-Checks in Video-Amplifier Signal Tracing

When the video amplifier appears to be distorting the camera signal, it is often helpful to make a follow-up square-wave test, as depicted in Fig. 4-20. It is good practice to disconnect the picture-detector diode in this test, because the diode would present a non-

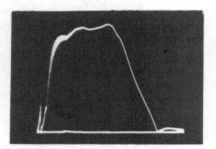

(B) Typical frequency-response curve.

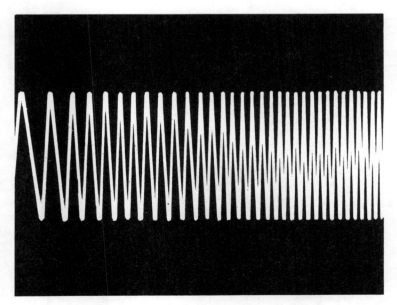

(C) Appearance of video-frequency sweep signal.

check of video-amplifier response.

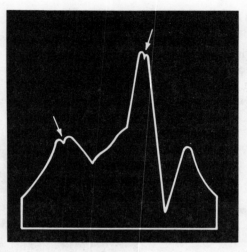

Fig. 4-24. Example of absorption markers on a video-frequency response curve (arrows point to markers).

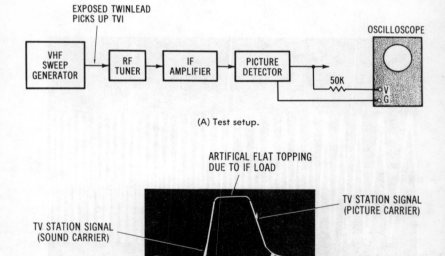

EXPOSED TWINLEAD
PICKS UP TVI

OSCILLOSCOPE

VHF
SWEEP
GENERATOR

RF
TUNER

IF
AMPLIFIER

PICTURE
DETECTOR

50K

V

G

(A) Test setup.

ARTIFICAL FLAT TOPPING
DUE TO IF LOAD

TV STATION SIGNAL
(PICTURE CARRIER)

TV STATION SIGNAL
(SOUND CARRIER)

TV STATION SIGNAL
(RESIDUAL VERTICAL-SYNC PULSE)

(B) Oscilloscope pattern.

Fig. 4-25. Television interference (tvi) may be picked up by high-gain test arrangements.

linear load to the generator output cable. A low-capacitance probe is applied at the output of the video amplifier. A 100-kHz square-wave test is generally regarded as basic, inasmuch as it shows whether the higher-frequency harmonics will be passed at normal amplitude and without objectionable phase shift. In other words, a 100-kHz square wave provides a critical test of video-amplifier transient response. The rise time of the output waveform should be in the range from 0.09 to 0.1 μs. Otherwise it is indicated that the video-amplifier bandwidth is subnormal. Basic square-wave distortions are depicted in Fig. 4-21. A small amount of overshoot and undershoot is often provided in video-amplifier design; the normal 100-kHz square-wave response of a high-performance video amplifier is shown in Fig. 4-22.

Rise-time measurements are made to best advantage with an oscilloscope that has a considerably faster rise time than the video amplifier; in such a case no correction factor is required. On the other hand, if the oscilloscope has a rise time in the same general "ball park" as the video amplifier, an accurate measurement requires that a correction factor be applied to the rise time that is displayed on the oscilloscope screen. This correction factor is called the *square root of the sum of the squares* law. For example, suppose that a video amplifier has a rise time of 0.1 μs, and the oscilloscope also has a rise time of 0.1 μs. Then the rise time of the square wave displayed on the oscilloscope screen will have a rise time of 0.14 μs, approximately. In other words, 0.1 squared is equal to 0.01, the sum of both squares is equal to 0.02, and the square root of 0.02 is equal to 0.14, approximately.

Video-Amplifier Frequency Response

When trouble symptoms throw suspicion on the frequency response or bandwidth of the video amplifier, it is helpful to make a sweep-frequency check, as shown in Fig. 4-23. An oscilloscope is connected to the output terminal of the video amplifier via a demodulator probe. The picture-detector diode is disconnected, as indicated in Fig. 4-23A, and a video-frequency sweep-and-marker signal is injected at the input of the video amplifier. The frequency-response curve for the amplifier will be displayed on the oscilloscope screen, as exemplified in Fig. 2-23B. The top of the curve should be reasonably flat, and the bandwidth should be at least 3.5 MHz. Note that either conventional beat markers or absorption markers may be provided by the generator. As a practical note, *when absorption markers are used, the demodulator probe must be capable of passing the rapid envelope*

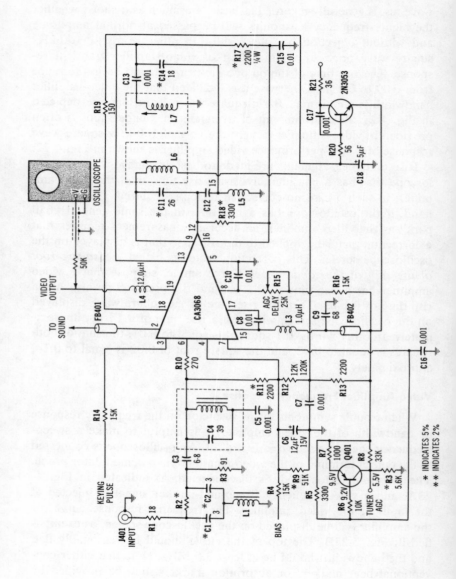

(A) Interconnections.

Fig. 4-26. Video-if configuration with

fluctuation; a demodulator probe such as shown in Fig. 3-13 is suitable, whereas a probe such as depicted in Fig. 3-11 tends to "wipe out" absorption markers.

Observe that the frequency response in Fig. 4-24 would be regarded as a trouble symptom; a highly peaked curve results in objectionable ringing ("circuit ghosts") when the video amplifier is passing signals with sudden transitions (white-to-black, or black-to-white). Common causes of frequency distortion in video amplifiers are open decoupling or bypass capacitors, substantially incorrect values of load resistors, and defective peaking coils. Note that the inductance values of peaking coils are often quite critical; if an error is made in replacement of a defective peaking coil, a "tough-dog" situation may occur that defies correction. Peaking coils ordinarily do not deteriorate unless exposed to moisture over considerable periods. However, peaking coils are subject to mechanical damage, such as charring from careless use of a soldering gun.

TELEVISION STATION INTERFERENCE

In high-gain oscilloscope test arrangements, television station (or other) interference (tvi) may be picked up unless suitable precautions are taken. For example, Fig. 4-25A shows a sweep-frequency test setup for checking the overall rf/if frequency-response curve. Even if the rabbit-ear antenna is removed from the rf tuner, there is an exposed length of twinlead that is capable of picking up tvi in areas that have high field strength. In turn, when the weak-signal frequency response of the receiver is checked, tvi is likely to appear in the oscilloscope pattern as a residual vertical-sync pulse, a station-signal picture carrier, and a station-signal sound carrier. Moreover, the tvi signal tends to overload the high-gain rf/if system, with the result that the overall frequency-response curve may be artificially flat-topped. To cope with tvi in this type of situation the exposed

STEP 1

1 Connect + 20 Volts to Appropiate Points on Board.
2 Connect Sweep Generator to Input.
3 Connect DC Bias Voltage to Appropiate Point on Board.
4 Adjust Sweep Generator for 10-Millivolt Input.
5 Adjust Bias Voltage for 5-Volt, Peak-to-Peak,Output.

(B) Sweep-alignment procedure.

STEP 2

1 Adjust L1 for Minimum Response at 47.25 MHz.
2 Adjust L2 for Maximum at 44.5 MHz.
3 Adjust L6, L7 for 3-MHz Bandwidth Centered at 44.5 MHz.

RCA CA3068 integrated circuit.

section of twinlead should be removed. In areas of unusually high field strength it may be necessary to move the receiver into a screened room while troubleshooting with the oscilloscope.

INTEGRATED-CIRCUIT IF SYSTEMS

Occasionally, the troubleshooter will encounter an integrated-circuit if amplifier, as exemplified in Fig. 4-26. All of the transistors and diodes for if amplification and detection are contained within the integrated circuit. Thus, when there is weak or no video output, the technician must determine whether the IC is defective, or whether an associated external component has failed. Note that if selectivity (tuning) is provided by the single-tuned input circuit with trap and by a double-tuned interstage circuit to the right of the IC. The resistive pad, R1, R2, and R3, terminates the link cable and isolates the cable effectively from the tuned input circuit. Resistor R1 feeds into a bridged-T trap for maximum attenuation of the adjacent sound carrier. Resistor R2, C1, and C2 are comparatively close-tolerance components; they provide a null at 47.25 MHz. Resistor R11 and the resistive input network determine the circuit bandwidth—a 3-dB bandwidth of 3 MHz with a center frequency of 44.5 MHz.

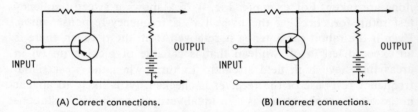

| (A) Correct connections. | (B) Incorrect connections. |

Fig. 4-27. Stage gain is reduced to approximately half of normal if the emitter and collector terminals are accidentally reversed.

Coils L5, L6, and L7 provide if interstage coupling, as in the analogous transistor-if configurations. A sweep-alignment check of the system may be made in the usual manner, as shown in the alignment procedure in Fig. 4-26. An input signal of 150 mV will normally produce full video output. In the event that a normal response curve cannot be obtained (3-MHz bandwidth centered at 44.5 MHz), or if excessive input signal voltage is required to obtain normal video output, the dc voltages specified in the receiver service data should be measured. Incorrect voltage values can provide helpful clues concerning faults in components or devices external to the IC, as opposed to defects within the IC. From a preliminary troubleshooting viewpoint the IC may be regarded as "a large transistor with 19 terminals." If troubleshooting tests indicate that the IC is defective, a substitution test is routinely made.

TRANSISTOR REPLACEMENT

As a practical note, be careful not to accidentally interchange emitter and collector leads (Fig. 4-27). Since the emitter and collector substances are doped differently and have different junction areas, the stage gain will typically be reduced to half if this error is made. In turn, the technician is confronted with a "tough-dog" situation—with the new transistor inserted (incorrectly) he or she is likely to waste a lot of time looking for trouble in the wrong places.

CHAPTER 5

Signal Tracing in the Sync Section

The sync section functions to keep the picture locked on the tv screen. Separate sync sections are utilized for black-and-white locking and for color locking. The horizontal- and vertical-sync sections separate the sync pulses from the composite video signal, step up the pulse amplitudes, process the pulses, and channel horizontal pulses into the afc section, and channel vertical pulses into the vertical-oscillator section.

OSCILLOSCOPE APPLICATION TECHNIQUES AND PATTERN EVALUATION

The plan of a simple black-and-white sync system is shown in Fig. 5-1. The composite video signal from the video amplifier is passed through a reverse-biased transistor which operates as a clipper and rejects the camera signal. Only the stripped sync pulses are passed into the sync amplifier. From the sync amplifier the stripped sync is applied to the integrator circuit and to the differentiator circuit. These circuits develop a vertical-sync "sawtooth" and horizontal-sync "spikes," respectively. The former waveform synchronizes the vertical-deflection oscillator, and the latter waveform synchronizes the horizontal-deflection oscillator (indirectly via the afc system).

In another widely used arrangement a sync-amplifier stage precedes the sync-separator stage, as exemplified in Fig. 5-2. In normal operation the sync amplifier provides a gain of 9, and the sync separator

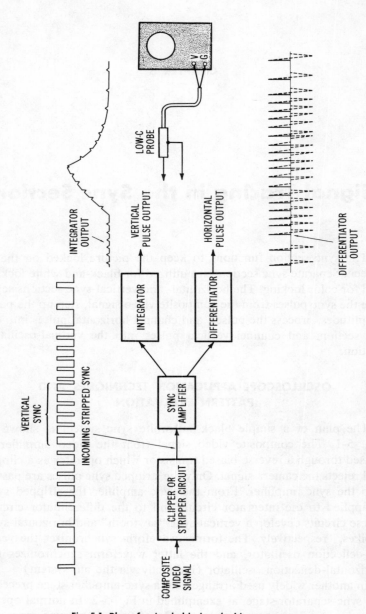

Fig. 5-1. Plan of a simple black-and-white sync system.

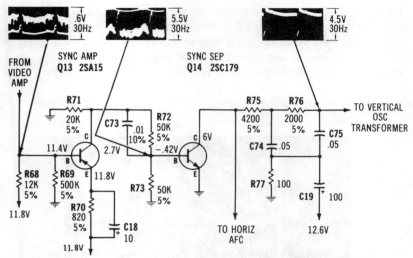

Fig. 5-2. The sync separator (clipper) is preceded by a sync amplifier in this design.

provides a loss of approximately 20 percent. Observe that the sync-amplifier transistor is biased into the nonlinear region of its operation, with the result that the camera signal is greatly compressed, and the sync pulses are extensively "stretched." In other words, partial sync separation occurs in the sync-amplifier stage. The output waveform

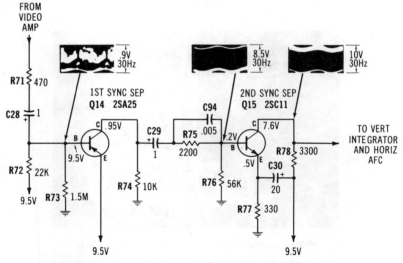

Fig. 5-3. A sync section that comprises two sync-separator stages.

from the sync separator contains only a trace of camera signal. Another widely used design of sync-section circuitry is shown in Fig. 5-3. Here, two sync-separator stages are employed. The first stage is biased at cutoff and normally provides a gain of approximately nine times. The second stage is reverse biased and provides a gain

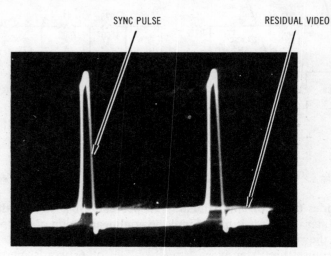

(A) Sync waveform with residual video signal.

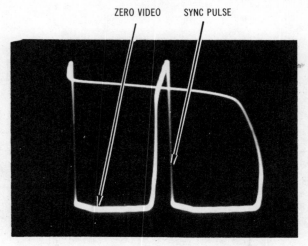

(B) Sync waveform with residual video signal removed.

Fig. 5-4. Action of two-stage sync separator.

of only 1.1. As a practical troubleshooting note, if abnormal wave-forms are observed in this type of configuration, the electrolytic capacitors should be checked first.

Stripping action in the first stage and in the second stage of a typical two-stage sync separator is shown in Fig. 5-4. Observe that the output waveform from the first stage has a greatly stretched horizontal-sync pulse, with a highly compressed video (camera signal) residue. Thus the first stage operates more as a clipper than as a sync amplifier. For optimum locking action it is desirable to remove the residual video component. Next, the output waveform from the first stage is passed through the second separator stage, with the result shown in Fig. 5-4B. Note that only the sync pulse appears in the output waveform—the residual video component has been clipped completely. In troubleshooting with the oscilloscope this processing of the signal is the chief consideration; if residual video signal happens to appear in the output waveform from the second stage, the technician would conclude that there is a component or device defect in the circuitry.

Kickback Component

With reference to Fig. 5-2, it is important to note that the waveform at the output of the integrator contains two signal components. There is the integrated vertical-sync component, and there is also the *vertical-oscillator kickback component*. These components are illustrated in Fig. 5-5. Note that if the picture is locked in sync vertically the integrated vertical-sync pulse is invisible—it is masked by the high-amplitude kickback pulse. However, if a split picture is being reproduced vertically, the integrated vertical-sync pulse is then displayed between the kickback pulses in the waveform on the oscilloscope screen. Or, if the picture is rolling vertically, the integrated vertical-sync pulse will "slide" through the waveform on the oscilloscope screen; that is, the oscilloscope locks on the high-amplitude kickback pulses. This is a practical point in troubleshooting with the oscilloscope, because the kickback pulse could otherwise be confused with the integrated vertical-sync pulse, thereby causing a "tough-dog" problem for the technician.

Noise Reduction

A *horizontal-sync channel should have a bandwidth of 135 kHz, from 15 kHz to 150 kHz.* This bandwidth is optimum for noise reduction and consequent freedom from erratic picture-pulling during weak-

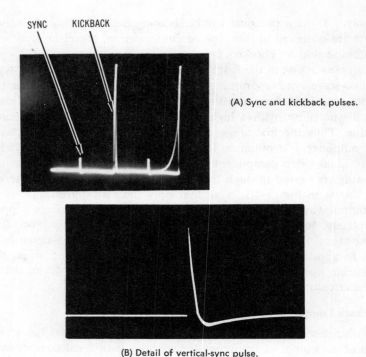

SYNC   KICKBACK

(A) Sync and kickback pulses.

(B) Detail of vertical-sync pulse.

Fig. 5-5. Integrated vertical-sync pulse appears between kickback pulses when reproduced picture is split vertically.

signal reception. Note that noise is seldom a problem in the vertical-sync channel, because the vertical integrator tends to absorb noise pulses. Most noise pulses are comparatively narrow and have high-frequency harmonics. Therefore the noise voltages can be reduced by utilizing a narrow-band horizontal-sync channel. The channel bandwidth cannot be reduced beyond 135 kHz, however, or the horizontal-sync pulses will also become objectionably attenuated. The bandwidth is controlled by RC values in the sync-separator circuitry. Observe the two-stage sync-separator configuration shown in Fig. 5-6. The C2-R6 network provides an upper frequency-response limit of 150 kHz. Note also that C6 and R12 provide a lower frequency-response limit of 50 Hz in the vertical-sync channel, to ensure rejection of possible low-frequency interference.

In addition to restricted sync-channel bandwidth the troubleshooter will find noise-switch devices included in sync-separator configurations to provide improved noise immunity. A noise switch is basically a

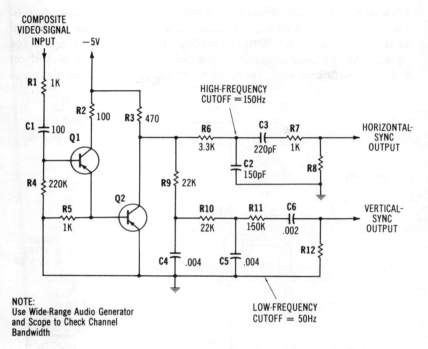

Fig. 5-6. A modern sync-separator arrangement that employs restricted bandwidth.

"hole-puncher" in the sync-signal channel. When a high-level noise pulse occurs, the noise-switch device briefly "opens" the sync-separator circuit so that passage of the high-level noise pulse is blocked. As soon as the noise pulse decays, the sync-separator circuit is enabled again. It is found in practice that less disturbance of picture synchronization results from "hole-punching" than if high-level noise pulses were permitted to pass. With reference to Fig. 5-7, the sync-separator transistor is controlled by the noise-switch transistor. Thus, when the signal-to-noise ratio is high, noise-switch transistor Q2 conducts and forms an emitter-return circuit for sync-separator transistor Q1. In turn, the sync separator operates as described previously. When a high-level noise pulse occurs, however, Q2 cuts off, and in turn the emitter circuit of Q1 is opened.

Observe in Fig. 5-7 that the sync-separator transistor is driven from the video amplifier, whereas the noise-switch transistor is driven from the picture detector. Diode X1 functions as a *noise separator;* this diode conducts only when the peak-to-peak noise amplitude exceeds the back-bias voltage developed by voltage divider R3–R4. Operation of the noise switch can be easily checked with an oscilloscope connected at the collector of Q2. With the receiver tuned to a weak signal (or to a vacant channel), noises pulses will normally be displayed on the oscilloscope screen. Then, with the receiver tuned to a normal

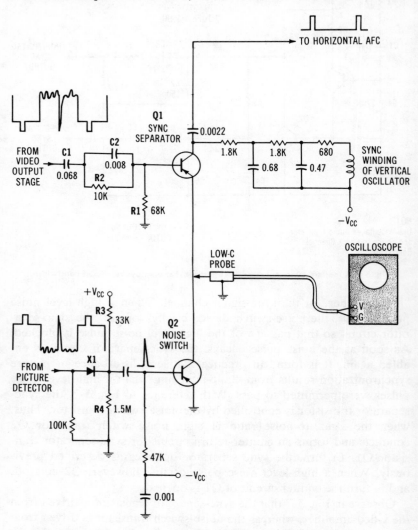

Fig. 5-7. A sync-separator transistor controlled by a noise-separator diode and a noise-switch transistor.

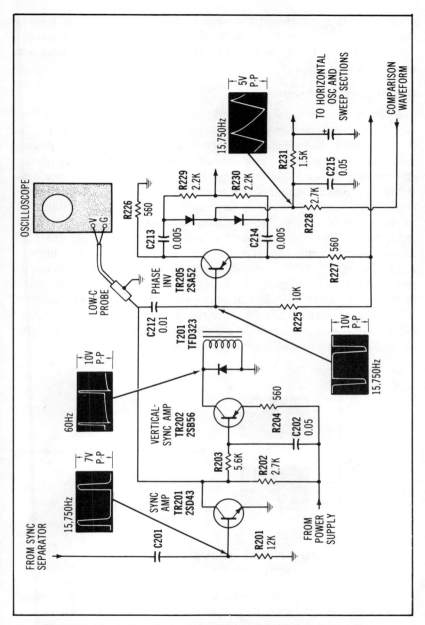

Fig. 5-8. A typical transistor sync amplifier and afc section, with separate vertical-sync amplifier.

149

or strong signal, no noise pulses are normally visible on the screen. If the noise-switch action is faulty, check for a defective noise-separator diode or a faulty noise-switch transistor. Leaky capacitors are also occasional culprits.

Operation of Sync Inverter

Many tv receivers utilize a phase-inverter stage, as shown in Fig. 5-8. The phase-inverter provides a push-pull sync-output signal to the afc section. Note that the collector-output signal from the phase-inverter transistor will have opposite polarity with respect to the emitter-output signal. As shown in Fig. 5-9 the collector waveform is "upside down" with respect to the emitter waveform. The two waveforms normally have the same peak-to-peak voltage. Note that in the example of Fig. 5-8 the integrator circuit is followed by a separate vertical-sync amplifier transistor with a protective diode in its collector circuit. The vertical-sync amplifier provides a higher-level sync signal. Note that the protective diode is polarized oppositely to the positive-going vertical-sync pulse; the diode functions to bypass the kickback pulse from the vertical oscillator and thereby to prevent damage to TR202. A horizontal-sync phase inverter provides double-ended or balanced afc action; tighter sync lock is thereby obtained than from single-ended or unbalanced afc circuits.

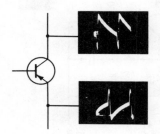

Fig. 5-9. Phase-inverter output waveforms.

COLOR-SYNC TROUBLESHOOTING WITH THE OSCILLOSCOPE

As shown in Fig. 5-10 the first subsection in the color-sync system is the burst takeoff and gating arrangement. The color burst on the back porch of the horizontal-sync pulse provides a frequency and phase reference for the 3.58-MHz color-subcarrier oscillator in the receiver. A gating pulse, derived from the horizontal-deflection section, is applied to the burst amplifier for separation of the color burst from the complete color signal. As indicated in Fig. 5-10B it is essential that the gating pulse be properly timed. A dual-trace oscilloscope can display both waveforms simultaneously and is advanta-

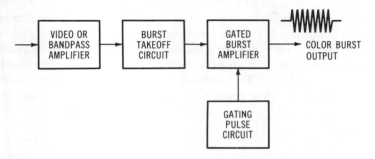

(A) Block diagram of subsection.

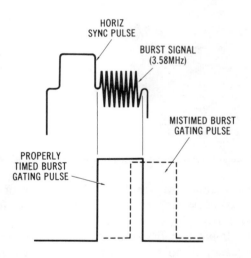

(B) Timing (check with dual-trace scope).

Fig. 5-10. First subsection in the color-sync system.

geous in checking gating-pulse timing. Note in passing that mistiming does not always result from a component or a device defect. The burst-gating pulse tends to lead or lag the color burst as the HORIZONTAL HOLD control is turned; thus a substantial misadjustment of the HORIZONTAL HOLD control will make it appear that there is a malfunction in the burst-gating circuitry.

Observe next that if the burst signal in Fig. 5-10 has subnormal amplitude, the output color burst may be too weak to lock the color picture properly. Attenuation of the burst signal can occur in the if amplifier, as depicted in Fig. 5-11. Note that the picture carrier and the color subcarrier normally fall at the 50 percent of maximum points on opposite sides of the if response curve. When the bandwidth of the if response curve is subnormal and its high-frequency response is deficient, the color subcarrier falls down lower on the curve and is attenuated accordingly. Therefore, if a waveform check of the burst amplifier shows that the color burst has considerably less peak-to-

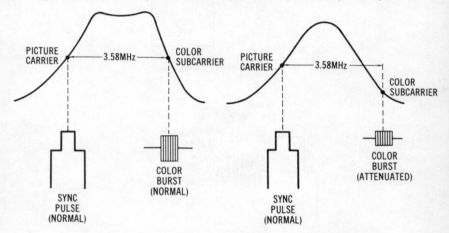

Fig. 5-11. Color burst can be attenuated by subnormal if bandwidth.

peak voltage than normal, the waveform at the output of the picture detector should be checked. Then, if the color burst is found to be weak at the detector output, a sweep-alignment check of the if amplifier is in order. Note that the color burst can also be attenuated by incorrect frequency response in the tuner, although this location is much less likely than the if amplifier; the if amplifier provides the major portion of the gain and selectivity in the picture channel.

In case that the waveform at the output of the picture detector is normal, the cause of a weak burst signal is most likely to be found in the burst amplifier circuitry. With reference to Fig. 5-12, the color burst can be attenuated by a weak gating pulse that does not drive Q12 into full conduction. Observe that Q12 is reverse biased 3.5 V.

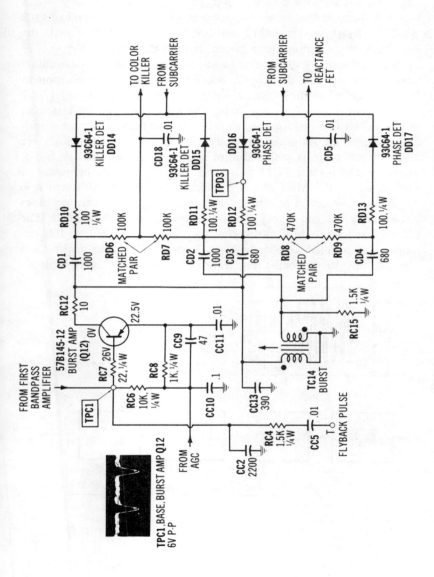

Fig. 5-12. Color burst can be attenuated by weak gating pulse.

153

In turn, the gating pulse must have a negative-peak voltage in excess of 3.5 V to drive Q12 into conduction. Although the gating pulse has a normal amplitude of 6 V p-p, its negative-peak voltage is less than 6 V because the pulse is ac-coupled into the base circuit of Q12 via CC5. Burst amplifier Q12 will not be driven to full gain if the amplitude of the gating pulse is substantially less than 6 V p-p.

In case that the gating pulse has normal amplitude but the color-burst output is weak, the troubleshooter should suspect a component or device defect in the burst-amplifier circuitry. For example, if Q12 develops collector-junction leakage, the stage gain will be less than normal. Off-value resistors can shift the bias voltage and cause amplifier malfunction. As noted previously, open or leaky capacitors are also common culprits in this kind of malfunction. Observe that burst-amplifier transformer TC14 is tunable. It has a normal bandpass of approximately 0.5 MHz and is peaked at 3.58 MHz. Although it is not as likely defective as other components, a burst transformer does occasionally become defective and cannot be correctly peaked. The color-burst amplitude is checked at the collector of Q12, in this example; its specified amplitude is 45 V p-p. This value might seem "impossible" to the beginning technician, inasmuch as the supply

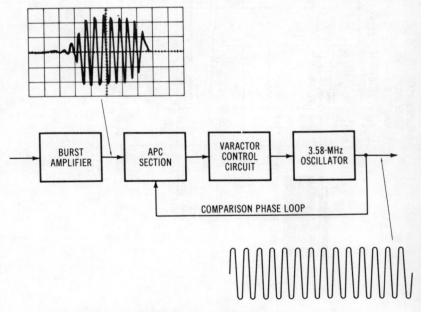

Fig. 5-13. Plan of the phase-locked-loop color-sync arrangement.

voltage is about half this value. However, it should be observed that TC14 provides a resonant load for the collector, and resonance involves a "flywheel" action and a resonant rise of ac voltage.

Phase Detector and Color-Sync Action

A widely used type of color-sync system utilizes a phase-locked loop (pll) configuration, as shown in Fig. 5-13. The output waveform from the burst amplifier is applied to a phase detector in the automatic phase-control (apc) section. The phase detector is basically a discriminator, as shown in Fig. 5-12. A correction voltage is normally developed, which operates on the varactor (or reactance FET) in

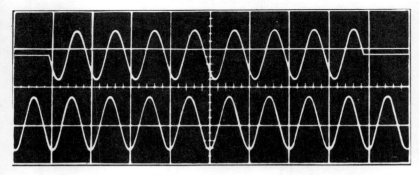

Fig. 5-14. Frequency and phase comparison with a dual-trace oscilloscope.

the control circuit to fine-tune the 3.58-MHz oscillator so that its subcarrier output is maintained exactly on frequency and in phase with the color burst. If a high-performance dual-trace scope is available, a frequency-and-phase comparison can be made as depicted in Fig. 5-14. It is advisable to use a color-bar generator as a signal source for the receiver so that a steady and controllable pattern is provided. Note that stability of pll action requires closely matched pairs of resistors and diodes in the phase-detector circuit.

Dual-Time-Constant Sync-Clipper Action

With reference to Fig. 5-15, a comparatively recent design of sync-clipper circuitry features a dual-time-constant circuit which is switched by a diode, X520. This configuration operates to clamp the sync clipper on successive horizontal-sync pulses and also to pass the vertical-sync block undistorted. The clipper circuit distinguishes between horizontal and vertical components in the composite

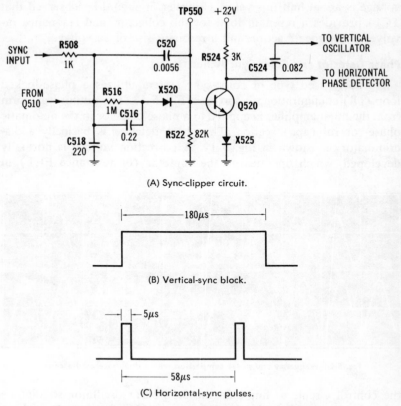

(A) Sync-clipper circuit.

(B) Vertical-sync block.

(C) Horizontal-sync pulses.

Fig. 5-15. Modern dual-time-constant design of sync-clipper circuit.

video signal. The vertical-sync block comprises six pulses with 27-μs duration; the total sync block has a duration of approximately 180 μs, and the interval between sync blocks is about 16,000 μs. The other component in the composite video signal is the horizontal-sync pulse, with approximately 5-μs duration and with a time interval between pulses of about 58 μs.

As noted previously, older designs of transistor sync circuits employ a large-value capacitor (0.2 to 1.0 μF) at the input of the circuit to couple vertical-sync pulses into the sync-clipper transistor without objectionable distortion. In this older type circuitry, however, the time constant of the coupling network is so large that the sync-clipper transistor may not always respond to some of the applied horizontal-sync pulses. The modern approach therefore involves a long-time-constant section for vertical-sync pulses, and a shorter-time-constant section for horizontal-sync pulses, with provision for automatically switching from one section to the other with each change from one signal component to the other. In Fig. 5-15A the section comprising C520, C516, R516, and X520 performs this automatic switching action.

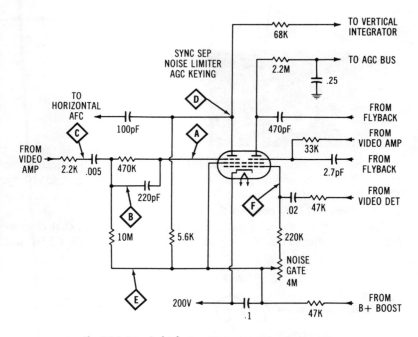

Fig. 5-16. A typical tube-type sync-separator arrangement.

Note that during the vertical-sync-block interval C516 charges up through R508, X520, and Q520. Also, C520 charges up at this time, with the voltage across it being equal to the voltage across C516 and X520. Then, after the vertical-sync block has passed, C520 and C516 start to discharge; capacitor C520 discharges through R522, and C516 discharges through R516. The time constant of C520–R522 is shorter than that of C516–R516; in turn, C520 discharges more rapidly than C516. Consequently, X520 becomes reverse biased between the vertical-sync-block intervals. In the time interval between vertical-sync blocks, there are approximately 262 horizontal-sync pulses accompanying the video information for one raster field. Be-

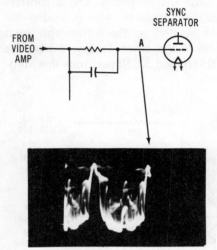

Fig. 5-17. Normal input to sync separator.

cause the signal path through C516 has been opened by the reverse bias on X520, these sync pulses are coupled to the base of Q520 by C520. In summary, C516 and X520 couple the vertical pulses into Q520, whereas C520 couples the horizontal pulses.

Otherwise this new design of sync clipper is similar to that of conventional designs. Thus the signal-developed bias across C516 and

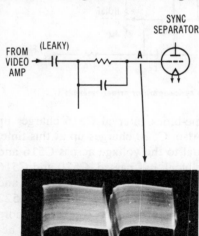

Fig. 5-18. Coupling capacitor leaky.

C520 biases the base of Q520 so that only the sync tips extend into the conduction region of the transistor. The camera signal is rejected, and only the sync tips appear in the collector circuit of the transistor. Note that when Q520 is driven into conduction by a sync pulse, the collector voltage falls nearly to zero; the transistor is driven into saturation. After the sync pulse passes, the transistor comes out of saturation and the collector current approaches zero as the base and emitter come to practically the same potential. Thus, between sync pulses the collector voltage on Q520 rises to almost 22 V.

SYNC TROUBLESHOOTING IN OLDER RECEIVERS

Although tube-type receivers are obsolescent, many are still in service. A typical and widely used sync-separator circuit used in these receivers is shown in Fig. 5-16. Sync separation, plus noise limiting, takes place in the left-hand section of the dual pentode; the right-hand section of the 6BU8 functions as an agc keyer. The normal signal input waveform to the sync separator is illustrated in Fig. 5-17. This signal is checked at point *A* with a low-capacitance probe. If this test point is checked on a vacant channel, a random-noise pattern normally appears. The signal amplitude at this point is approximately 30 V p-p. Although picture lock will normally be

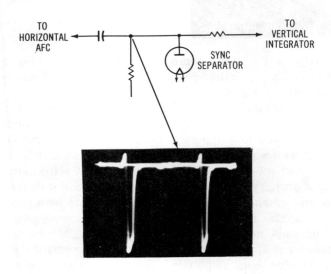

Fig. 5-19. Normal separated sync signal.

maintained at lower amplitudes, substantial signal attenuation results in unstable sync action or complete loss of locking action. In case that the signal amplitude is found to be low at point A, a follow-up check should be made at points B and C of Fig. 5-16. Note that an open capacitor in the grid circuit will cause excessive signal attenuation with waveform distortion. Or, if the 0.005-μF capacitor becomes leaky, the dc grid bias will be shifted, and the waveform at point A will become blurred and attenuated (Fig. 5-18).

In the event that the signal waveform is found to be normal at point A, the next checkpoint is at D; the normal waveform at the plate of the separator tube consists of cleanly stripped sync pulses, as shown in Fig. 5-19, with only a slight trace of residual video signal along the top of the waveform. If the waveform is found to be normal at point D, horizontal-locking trouble is then logically sought in the afc or horizontal-oscillator section. Similarly, vertical locking trouble will logically be looked for in the vertical integrator or in the vertical oscillator section. On the other hand, if the waveform at point D is

Fig. 5-20. Unsatisfactory sync separation.

not normal and if it shows appreciable residual signal (Fig. 5-20), there is probably a defective component in the plate circuit. For example, the 100-pF coupling capacitor may be leaky.

Faulty sync separation can also be caused by a defect in the cathode circuit. To trace this signal, check the waveform at point E. In normal operation a low-amplitude video signal on the order of 5 V p-p is found at this point. Little or no signal at point E indicates that the 0.1-μF cathode bypass capacitor may be leaky. However, if this capacitor is open, sync-separation action is not as seriously disturbed. The first grid in the tube (Fig. 5-16) is common to both sections of the dual pentode. This first grid operates in the noise-gate circuit. A low-amplitude, negative-going signal (about 0.2 V p-p) from the video detector is applied at point F, and has the waveform depicted in Fig. 5-21. The signal amplitude is normally too low to affect sync-separator action unless a high-level noise pulse arrives. In such a

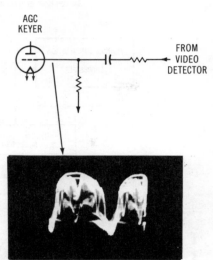

FROM
VIDEO
DETECTOR

Fig. 5-21. Negative-going noise gate.

case the high negative-peak voltage of the noise pulse cuts off the tube for the duration of the noise pulse. Thus a "hole" is punched in the separator output signal; in practice, sync stability is far better than when a noise-gate circuit is not utilized.

In case that the 0.02-μF coupling capacitor to the noise-gate grid is open, the negative-going video signal obviously will not feed into the grid circuit, and the noise gate then becomes inoperative. This

SYNC
SEPARATOR

FROM
VIDEO
AMP

A

Fig. 5-22. Distortion of video signal, caused
by loss of horizontal sync.

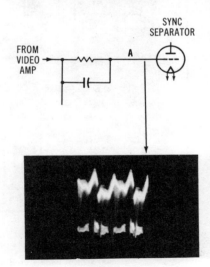

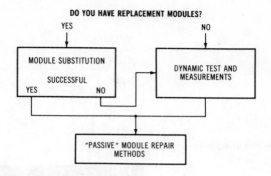

(A) Module troubleshooting scheme.

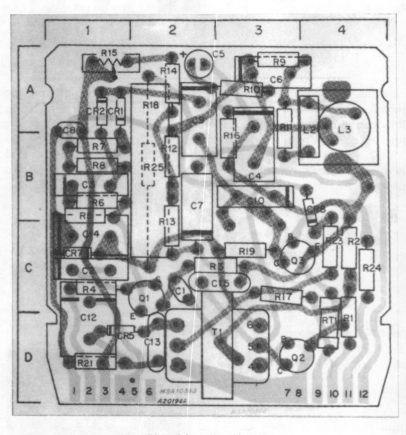

(B) Module pc board layout.

Fig. 5-23. A widely used horizontal-oscillator module.

failure is of no consequence during strong-signal reception. On weak and noisy channels, however, horizontal-sync lock becomes less stable, because high-level noise pulses then feed through to the afc circuit. However, if the 0.02-μF coupling capacitor becomes leaky, the picture will disappear and no video signal will be found at point F. What happens in this situation is that dc voltage bleeds through to the video detector and "kills" the video signal, unless the noise-gate control is set to a high-resistance point.

The preceding review of sync-separator action is typical of tube-type receivers. Although numerous variations in circuit details are employed, the general principles of operation remain the same. Thus the sync separator always functions to strip the sync tips from the composite video signal, and it prevents passage of video signal. In the event that a noise gate is included, it functions to punch a "hole" in the separated sync signal for the duration of a high-level noise pulse. In each case the troubleshooter should consult the receiver service data to determine the normal waveforms and their peak-to-peak voltages.

A practical note of caution: Normal waveforms are sometimes distorted because of reflected abnormal waveforms from the horizontal-afc or horizontal-oscillator circuits. Also, receivers that have keyed agc may generate spurious ac pulses on the agc line whenever the picture loses horizontal-sync lock. This spurious pulse train can "chop up" the video-signal input to the sync separator, as shown in Fig. 5-22. To avoid being misled in this situation the technician should clamp the agc line with a bias box or battery before checking waveforms in the sync section.

NOTE ON MODULAR CONSTRUCTION

Many modern receivers employ modular construction, as exemplified in Fig. 5-23. An advantage of this design is that when a trouble symptom throws suspicion on a particular module, a replacement module can be quickly plugged in for a substitution test. Consequently the down time during a repair job can often be reduced considerably. A defective module can be repaired at any convenient time subsequently. Note that modular construction does not change any of the principles that have been described in troubleshooting with the oscilloscope. The basic circuitry is the same, and troubleshooting waveforms is the same, regardless of mechanical design. Thus, whether you are troubleshooting an older-model chassis with point-to-point wiring, or whether you are coping with modular printed-circuit arrangements, you will use the same troubleshooting procedures.

CHAPTER 6

Troubleshooting the AFC and Horizontal-Oscillator Section

With reference to Fig. 6-1, the afc section has two signal inputs: one from the sync separator (sync phase inverter) and the other from the horizontal oscillator. The pulses from the phase inverter are applied to one end of the afc diodes, and the sawtooth comparison waveform from the horizontal-driver circuit is applied to the other end of the afc diodes. Signal tracing through the horizontal-afc section is accomplished with a low-capacitance probe. It is helpful to drive the receiver from a pattern generator so that a stable and controllable test signal is processed by the receiver system.

OVERVIEW OF HORIZONTAL-AFC CIRCUIT ACTION AND WAVEFORMS

The configuration shown in Fig. 6-1 is called a *balanced afc network*. A comparison waveform from the collector of driver transistor Q21 is fed through L17 and C39 to the center of the balanced-diode afc circuit. *Incorrect waveform amplitudes and distorted waveshapes in this section are often caused by unmatched afc diodes.*

Note in Fig. 6-1 that C96, C38, R123, R124, C97, and C98 form a coupling, filtering, and antihunt network. An antihunt network provides system operation with an input-output phase relation which prevents self-oscillation. In the event of off-frequency operation it is

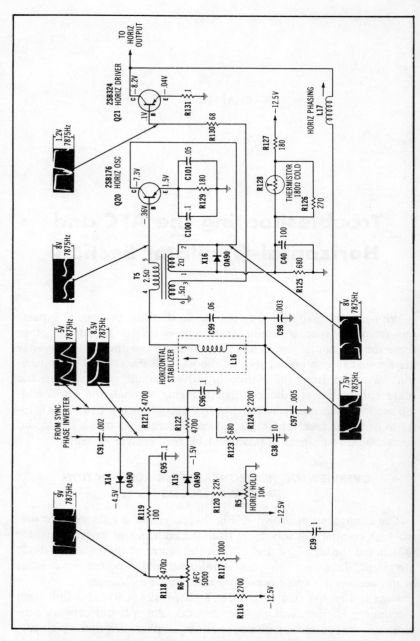

Fig. 6-1. A solid-state horizontal-afc, oscillator, and driver configuration.

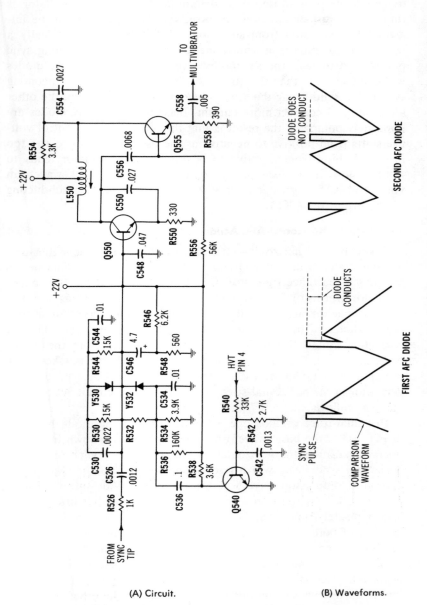

(A) Circuit.   (B) Waveforms.

Fig. 6-2. A widely used horizontal-afc and oscillator arrangement.

often helpful to measure the horizontal-oscillator frequency with a triggered-sweep oscilloscope to determine whether this oscillator is running too fast or too slow. In turn, some components can be immediately eliminated from suspicion. The afc circuit is basically a wave-timing comparison arrangement. That is, if the incoming sync pulses tend to lead the sawtooth waveform, one of the afc diodes will conduct more than the other. On the other hand, if the incoming sync pulses tend to lag the sawtooth comparison waveform, the other afc diode will conduct more current. Inasmuch as the afc diodes are oppositely polarized, the result is that the afc output dc control voltage shifts from positive to negative as the oscillator tends to run too fast, and shifts from negative to positive as the oscillator tends to run too slow. In the example of Fig. 6-1 the peak-to-peak waveform voltage at X15 is greater than at X14, and thus X15 is conducting more current than X14.

Single-Ended Horizontal-AFC Action

A widely used horizontal-afc and oscillator arrangement that uses a single-ended sync input circuit is shown in Fig. 6-2. This design includes a reactance transistor, Q550, in addition to the horizontal-oscillator transistor, Q555. Thus Q550 acts effectively as an electronic capacitor across oscillator coil L550, and serves to correct any tendency of the oscillator to run too fast or too slow. The afc correction voltage is applied to the base of Q550 from the phase-detector circuit. Output from the horizontal oscillator is taken from the emitter of Q555, and it triggers the multivibrator type of pulse generator in the horizontal-sweep section. The afc circuit action takes place as follows.

If a dual-trace oscilloscope is connected into the diode branches in Fig. 6-2A, the basic waveforms are displayed as shown in normal operation. Unless the horizontal oscillator has lost synchronism, the horizontal-sync pulse will arrive during the flyback interval in the sawtooth comparison waveform. Since the two diodes are oppositely polarized, the comparison waveforms in the two circuit branches are also oppositely polarized. The sync pulse, however, is applied to the cathode of both of the afc diodes. In turn, the sync pulse "rides on" the leading edge of one comparison waveform, but "rides on" the trailing edge of the other comparison waveform. Accordingly, if the horizontal oscillator tends to run too fast, the sync pulse will ride higher on one waveform, but will ride lower on the other waveform. The end result is that one afc diode will conduct more than the other

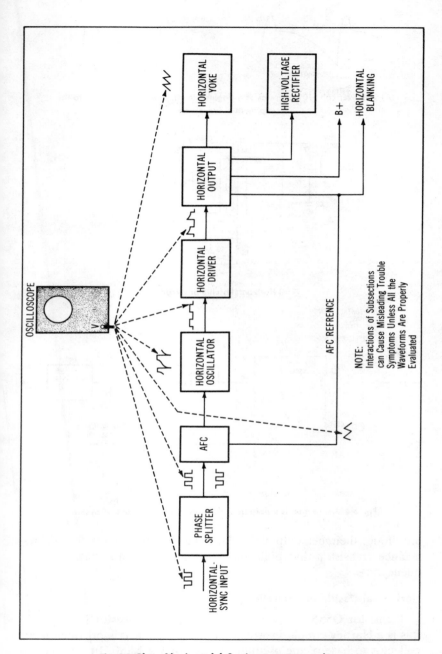

Fig. 6-3. Plan of horizontal-deflection system operation.

169

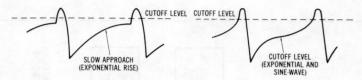

(A) Stabilizer coil L101 operates as a ringing circuit, to add a sine wave to the exponential waveform.

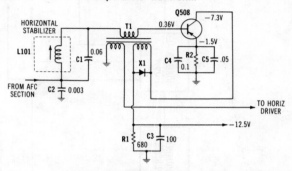

(B) Horizontal-oscillator circuit.

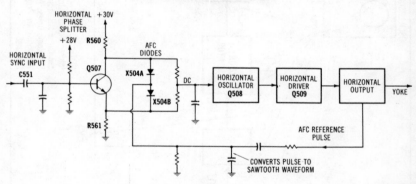

(C) Horizontal-deflection system.

Fig. 6-4. Afc section is a portion of the servo (feedback control) system.

afc diode; their net output develops the control voltage for the reactance transistor that pulls the horizontal oscillator back on frequency.

Horizontal-Oscillator Operation

Transistor Q555 is the horizontal-oscillator transistor in Fig. 6-2; this is a Hartley circuit, in which inductive feedback is provided from collector to base of the oscillator transistor via L550. In normal op-

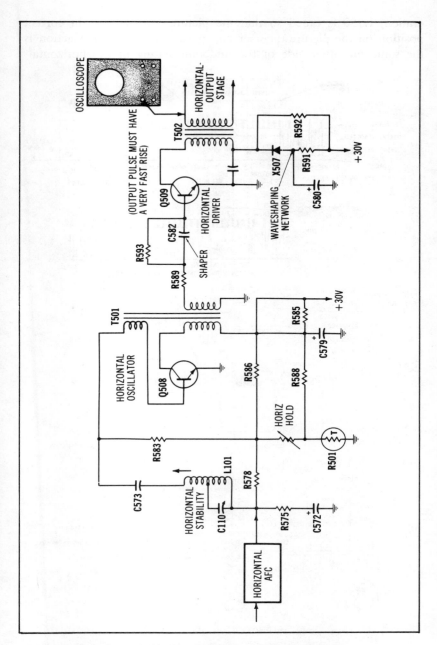

Fig. 6-5. Typical horizontal-oscillator and driver configuration.

eration, L550 is tuned so that the picture "freewheels" in correct position on the picture-tube screen, or so that the locking action is the same on either side of the midpoint setting of the horizontal-

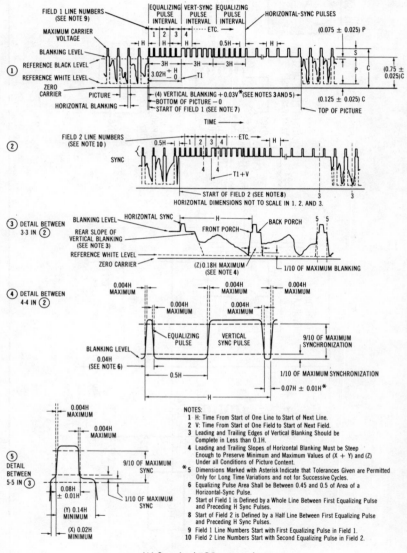

(A) Standard FCC sync pulses.

Fig. 6-6. Sync

hold control. The Hartley oscillator is essentially a sine-wave source, although its output is not a pure sine wave. Normal waveforms and peak-to-peak voltages are usually specified in the receiver service data. In case that the circuit fails to oscillate, the transistor is a likely suspect—for example, it may be found to have substantial collector-junction leakage. If the transistor is not defective, however, capacitors in the oscillator circuit should be checked for leakage or open circuits. Although off-value resistors or a defective oscillator transformer can cause oscillator failure, these faults are the least likely.

Horizontal-Deflection System

Troubleshooting with the oscilloscope is facilitated by an understanding of system operation. With reference to Fig. 6-3, each subsection performs a specific function, and the final result is synchronized horizontal scanning of the picture-tube screen. Horizontal-sync pulses, stripped from the composite video signal, are applied to the phase splitter, which in turn develops a push-pull pulse output. This double-ended output consists of horizontal-sync pulses that are equal in amplitude but opposite in polarity. This dual waveform output is used as a timing reference by the afc circuit, where the phase and frequency of the horizontal-output voltage is automatically checked against that of the reference pulses. Note in Fig. 6-3 that if the horizontal-blanking circuit in the horizontal-output system becomes defective, there will be *apparent* trouble in the afc section. Any confusion that might arise in this regard can be avoided by checking the afc reference waveform to localize the trouble source.

As noted previously the afc circuit normally develops an error voltage which is proportional to the phase or frequency difference between the incoming sync pulses and the horizontal-deflection waveform. Observe that in some receivers the dc error voltage may be applied directly to the base of the horizontal-oscillator transistor,

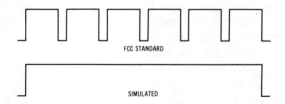

FCC STANDARD

SIMULATED

(B) Simplified sync generator may produce a vertical pulse without serrations.

waveforms.

as shown in Fig. 6-4. Frequency control is provided in this design by changes in the discharge time of the base-emitter capacitors that are charged by development of signal overdrive (self-developed bias). The error voltage determines the precise time at which Q508 comes out of cutoff.

With reference to Fig. 6-4, stripped sync pulses are applied via C551 to the phase splitter, Q507. The horizontal oscillator transistor, Q508, operates as a *blocking oscillator,* with refinements to improve operating stability. The power supply is regulated, and horizontal stabilizer coil L101 shapes the waveform output from the blocking oscillator. Transistor Q508 operates between saturation and cutoff,

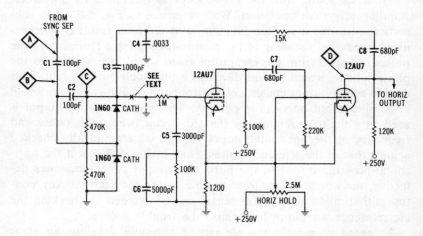

Fig. 6-7. Typical tube-type horizontal-oscillator and afc configuration.

with the result that the output approximates a rectangular pulse.

Because the afc, horizontal-oscillator, and horizontal-sweep sections operate within a servo (feedback control) loop, input/output phase relations are basic stability factors. From the viewpoint of the troubleshooter this circuit action has a potential for *system self-oscillation* should component defects produce a significant change of the input/output phase relation. The horizontal-driver stage (not shown in Fig. 6-4) amplifies the rectangular pulse output from Q508 and shapes it into a fast-rise pulse suitable for driving the horizontal-output stage. A typical horizontal-oscillator and driver configuration is depicted in Fig. 6-5. The driver circuit action is basically that of a clipper-amplifier, which speeds up the pulse rise time by selecting

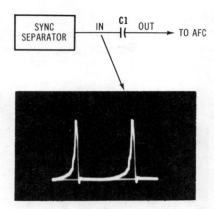

Fig. 6-8. Normal input waveform obtained
from sync separator.

a central "slice" of the input pulse and then amplifying it. *When the horizontal-output system does not operate normally, and particularly if the horizontal-output transistor overheats, the troubleshooter should start by checking the rise time of the output pulse from the horizontal-driver stage.*

PATTERN-GENERATOR SYNC WAVEFORMS

High-performance pattern generators provide precise horizontal- and vertical-sync pulses. Economy-type generators may supply non-standard sync, with various simulations of true sync pulses. These departures from standard FCC waveforms may or may not be of practical significance in troubleshooting with the oscilloscope. The technician should be aware of these factors, however, so that he or she will not be confused in critical test situations. As shown in Fig.

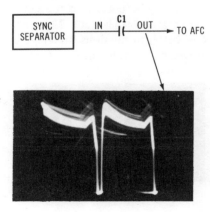

Fig. 6-9. Normal waveform at the output
end of the coupling capacitor.

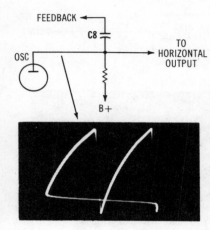

FEEDBACK ◄

C8

OSC

TO
HORIZONTAL
OUTPUT

B+

Fig. 6-10. Normal sawtooth waveform from
horizontal-oscillator circuit.

6-6 the standard FCC sync waveform provides for interlaced scanning; there is a difference of half a scanning line between the last horizontal pulse and the first equalizing pulse in even fields, compared with odd fields. Many economy-type generators do not provide for interlaced scanning. Note also in Fig. 6-6 that the standard FCC sync waveform has a serrated vertical-sync pulse; the sync pulse contains five serrations. Simplified pattern generators often supply a vertical-sync pulse without serrations. In consequence, the receiver under test tends to lose horizontal-sync lock during passage of the vertical-sync pulses.

Sync pulses that were originally standard may also undergo deterioration in the course of network transmission. Pulses sometimes become distorted and may also be substantially attenuated. Hum voltage may gain entry into the network signal so that the sync-tip level varies at a 60-Hz rate. Noise voltages may also contaminate some network transmissions. In fringe areas network signals become further subjected to noise interference during propagation from transmitter

Fig. 6-11. Waveform from horizontal-oscillator circuit when coupling capacitor is leaky.

to receiver. Thus, in some situations, the technician may find that an off-the-air signal is less reliable than the signal from an economy-type pattern generator.

TROUBLESHOOTING TUBE-TYPE AFC AND HORIZONTAL-OSCILLATOR CIRCUITS

Although tube-type receivers have been displaced to a considerable extent by solid-state receivers, the tv technician must still cope with tube circuitry. Referring to Fig. 6-7, semiconductor diodes are commonly utilized in the afc section, and the horizontal oscillator operates in a cathode-coupled multivibrator configuration. Signal tracing starts with a check of the input signal from the sync separator at point A. In normal operation a waveform is found at point A, as

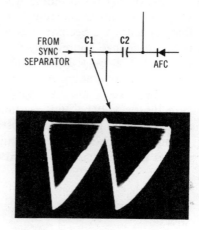

Fig. 6-12. Waveform at point B when C1 is open.

shown in Fig. 6-8. If the waveform is seriously distorted, or absent, the trouble will be found in the sync separator section and not in the afc circuit. On the other hand, if the sync separator is supplying a normal waveform, the next check should be made at point B of Fig. 6-7. Note that although the same waveform might be anticipated at points A and B, this is not the case. The reactance of the 100-pF coupling capacitor causes the waveshapes to be different—the waveform at B becomes mixed to some extent with a sawtooth component from the horizontal oscillator (Fig. 6-9). If the horizontal oscillator is inoperative, the same waveform will then appear at points A and B.

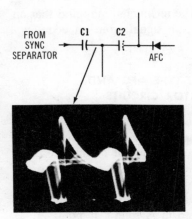

FROM
SYNC
SEPARATOR

C1 C2

AFC

Fig. 6-13. Waveform at point B when C2 is open.

Observe that the sawtooth waveform from the horizontal oscillator (Fig. 6-10) enters the afc circuit at point *C* in Fig. 6-7. But suppose that capacitor C2 becomes leaky. In this situation the waveform does not then change appreciably in amplitude, but becomes distorted, as shown in Fig. 6-11. Horizontal locking action is unstable under this condition of operation. As a general procedural rule, when waveform tests throw suspicion on a particular circuit, the next step is to measure the dc voltage and resistance values in the circuit. Capacitors are commonly checked by substitution, or they may be tested on a capacitor checker. Note particularly in Fig. 6-7 that the 1N60 afc diodes may become unbalanced or defective, and in turn cause sync

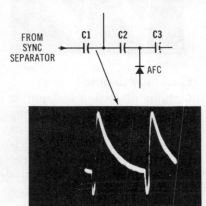

FROM
SYNC
SEPARATOR

C1 C2 C3

AFC

Fig. 6-14. Waveform at point B when C3 is open.

malfunction. Diodes are ordinarily checked by substitution, or they may be tested for front-to-back ratio with an ohmmeter. Matched diodes will have virtually the same front-to-back ratio.

Oscillator or AFC?

In some trouble situations the receiver is apparently out of sync lock because the horizontal oscillator is operating so far off frequency that the afc circuit cannot pull it into sync. Inasmuch as the cause of off-frequency operation could be either in the oscillator or in the afc section, a test method is desirable to make this determination. With reference to Fig. 6-7 the 1-megohm isolating resistor may be disconnected as indicated at X; the disconnected end of the resistor is then grounded as indicated by the dotted line. Then, if the trouble is in the afc circuit, it will be possible to freewheel the picture into

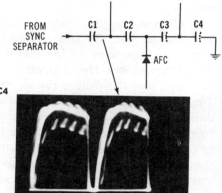

Fig. 6-15. Waveform at point B when C4 is open.

horizontal sync (at least momentarily) by critical adjustment of the HORIZONTAL HOLD control. On the other hand, if the trouble is in the oscillator circuit, it will not be possible to frame the picture by adjustment of the HORIZONTAL HOLD control.

The preceding test is based on the functional principle that the oscillator normally operates at approximately 15,750 Hz when the afc voltage is zero. Thus the control voltage is set to zero by means of the test ground connection to determine whether the oscillator will then operate on frequency. In the event that C1 becomes open, horizontal locking action is nearly lost and adjustment of the HORIZONTAL HOLD control is extremely touchy. In this situation the waveform at point B in Fig. 6-7 does not exhibit a prominent sync-pulse component; instead, it then appears as a distorted sawtooth (Fig. 6-12). In turn, the sawtooth amplitude is less than that of the normal waveform, because the pulse component is missing.

Next, if C2 is open, the waveform at point *B* will become distorted, as shown in Fig. 6-13. Although it might be anticipated that an open capacitor would decrease the waveform amplitude, that is not the case in this situation. The diode response changes when the capacitor becomes open, with the result that the waveform amplitude approximately doubles. Note that stability of sync lock is not greatly affected on strong channels, but sync lock becomes unstable on weak-channel reception. Note that the waveform shown in Fig. 6-13 (as in the case of the previous waveforms) is displayed with the HORIZONTAL HOLD control carefully adjusted to frame the picture in horizontal-sync lock. This is an important consideration, because the afc waveforms often become greatly changed and blurred if the picture is out of horizontal sync.

If C3 opens up, the distorted waveform shown in Fig. 6-14 will be found at point *B*. This waveform has about double the normal amplitude because of the increase in circuit impedance when C3 has no loading action. The comparison waveform (oscillator sawtooth) is absent, and the picture cannot be framed horizontally unless the 1-megohm resistor is disconnected and grounded as shown by the dotted lines in Fig. 6-7. Then the picture can be freewheeled into frame by careful adjustment of the HORIZONTAL HOLD control. Next, in case that C4 opens up, a distinctive form of distortion appears in the pattern at test point *B,* as shown in Fig. 6-15. This distorted waveform has an amplitude that is several times greater than normal. Note that when C4 is open its intended bypassing or attenuating action is defeated, with the result that the sawtooth comparison wave is displayed at substantially greater amplitude than normal.

CHAPTER 7

Waveform Tests in the Horizontal-Sweep Section

A standard SCR horizontal-deflection circuit has two silicon controlled rectifiers (SCRs) which act as bidirectional switches in combination with diodes, as shown in Fig. 7-1. Note that there is normally a peak power of 1200 volt-amperes in the yoke, as indicated in Fig. 7-2.

TROUBLESHOOTING THE SCR HORIZONTAL-SWEEP SYSTEM

With reference to Fig. 7-1 the trace-switch SCR, SCR_T, and diode X_T switch the deflection current through the yoke winding L_Y during the forward-trace interval. On the other hand, the retrace-switch SCR or commutating SCR, SCR_C, and diode X_C provide commutating-switch action for the flyback interval. At the beginning of the forward scan, trace-switch diode X_T conducts the yoke current that was stored as a magnetic field during the latter half of the previous scanning interval. This trace-switch diode conducts a linearly decreasing current ramp; as the current falls to zero the first half of the forward-scan interval is completed.

Just before the ramp reaches zero, the trace-switch SCR, SCR_T, is readied for conduction by the application of a positive pulse to its gate from resistor R_G. The yoke-current ramp then crosses the zero level from negative to positive, whereupon the circuit current transfers from trace-switch diode X_T to trace-switch SCR SCR$_T$, as shown in Fig. 7-3. Then capacitor C_Y begins to discharge via trace switch SCR$_T$ to drive yoke current through winding L_Y during the second half of the forward-scan interval. Note that in normal operation the voltage across C_Y changes but slightly during the forward-scan/retrace sequence. This virtually constant-voltage source drives a linearly rising current through the yoke winding.

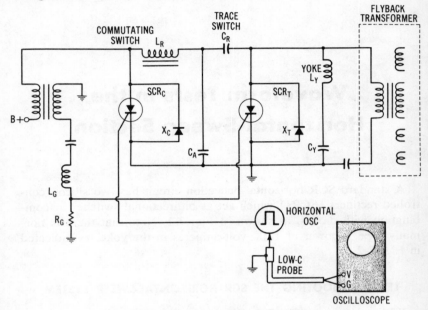

Fig. 7-1. Basic SCR horizontal-deflection circuit.

Commutating-Switch Action

As the end of the forward scan is approached, the commutating-switch SCR, SCR$_C$, is gated into conduction by a pulse from the horizontal oscillator. At this time capacitor C_R is enabled to discharge a current pulse through L_R via the trace and commutating SCRs. This discharge current is termed the *commutating pulse;* its peak-to-peak value increases until it is greater than the yoke current, whereupon trace diode X_T starts to conduct. This conduction by X_T reverse biases the trace SCR long enough that it cuts off. *Normal oscilloscope waveforms that accompany this sequence of circuit actions are shown in Fig. 7-4.* Note that as the commutating pulse falls to a value lower

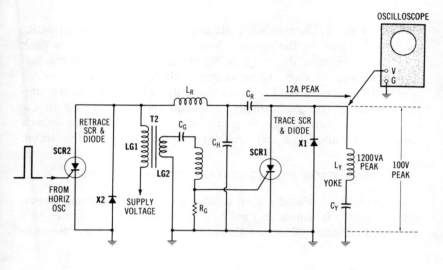

(A) Configuration.

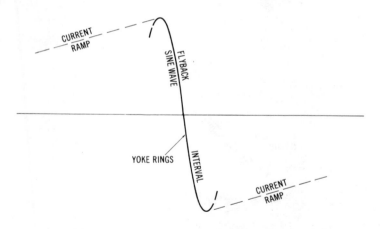

(B) Retrace interval is one-half of a 70-kHz sine wave.

Fig. 7-2. SCR horizontal-sweep arrangement.

than the amplitude of the yoke current, X_T starts to conduct; thereby the magnetic energy stored in the yoke winding produces a current which charges retrace capacitors C_A and C_R over the first half of the forward-scanning interval.

This charge "rings" back into the yoke over the second half of the forward-scanning interval (see Fig. 7-2B). The ringing circuit is completed via X_C, and an adequate time lapse is provided for SCR_C to cut off. Then, as the current through the yoke reaches its peak value in the negative direction, SCR_T begins to conduct and the forward-scan interval starts. Observe that while the commutating switch is closed, input inductor L_G is automatically switched across the supply-voltage source. Accordingly, magnetic energy stores up in the inductor. This stored magnetic energy serves to charge up the flyback capacitors C_A and C_R, thereby replenishing the I^2R loss in the circuit.

TRANSISTOR HORIZONTAL-OUTPUT ARRANGEMENT

A standard horizontal-output configuration that uses a pnp power-type transistor is shown in Fig. 7-5. Transistor Q102 operates in a grounded-collector (emitter-follower) circuit. The load for the emit-

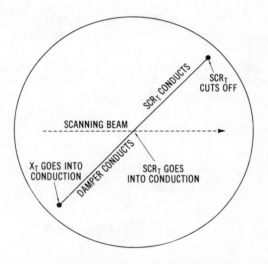

Fig. 7-3. Sequence of circuit events during the forward scanning interval.

ter consists of the high-voltage transformer, T102, the yoke winding, and the damper X105. The output stage starts to conduct at approximately the center point of the forward scanning interval, and it produces a linear current rise for the remainder of the forward scan. Then, at the end of the current rise, Q102 is suddenly cut off by the input waveform from the driver stage. At this time the flyback pulse is generated; the flyback produces retrace and drives damper X105 into conduction. Damper conduction continues for the first half of the forward-scanning interval, and decays to zero at approximately the center of the scanning interval. The cycle then repeats. Fig. 7-6 shows the waveforms that are involved in this horizontal-output circuit action.

To protect the output transistor from excessive current, a current-limiter transistor (Q108 in Fig. 7-5) is included. Also, diode X107

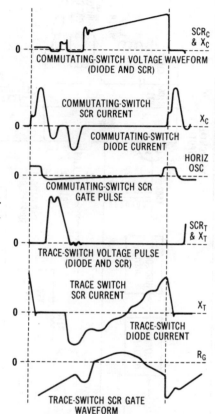

SCR$_C$ & X$_C$

COMMUTATING-SWITCH VOLTAGE WAVEFORM
(DIODE AND SCR)

COMMUTATING-SWITCH
SCR CURRENT

X$_C$

COMMUTATING-SWITCH
DIODE CURRENT

HORIZ OSC

COMMUTATING-SWITCH SCR
GATE PULSE

Fig. 7-4. Normal voltage and current waveforms in the SCR horizontal-sweep system.

SCR$_T$ & X$_T$

TRACE-SWITCH VOLTAGE PULSE
(DIODE AND SCR)

TRACE SWITCH
SCR CURRENT

X$_T$

TRACE-SWITCH
DIODE CURRENT

R$_G$

TRACE-SWITCH SCR GATE
WAVEFORM

protects the current-limiter transistor from breakdown due to spike voltages. The picture width is adjusted by means of the width coil L107, which is connected in series with the horizontal-yoke windings. Note that the adjustment of L107 also affects the efficiency of yoke operation. High-voltage pulses are generated as a result of the flyback pulse that occurs during retrace time. This flyback pulse is stepped

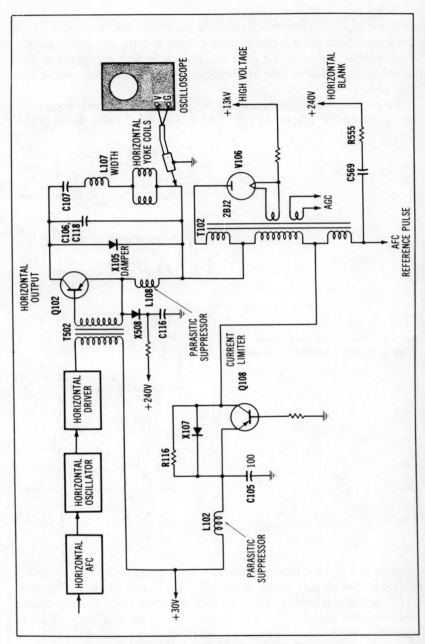

Fig. 7-5. A widely used horizontal-output transistor arrangement.

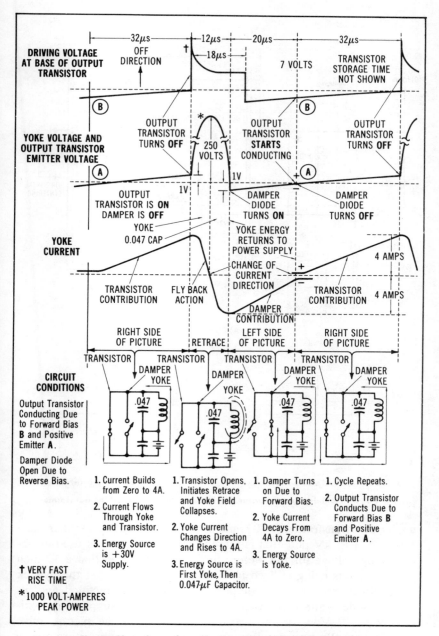

Fig. 7-6. Normal waveforms for the configuration of Fig. 7-5.

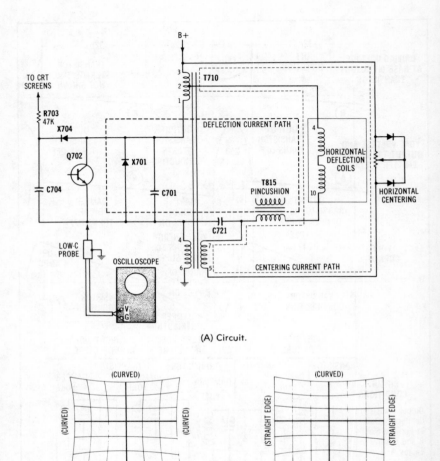

(A) Circuit.

(B) Top-bottom and side pincushioning.

(C) Top and bottom pincushioning.

Fig. 7-7. Horizontal-output transistor circuit that includes a pincushion transformer.

up by autotransformer T102 and is then rectified by rectifier tube V106. A +13-kV potential is thereby obtained, which is used as accelerating voltage for the picture tube. (It is assumed that a 12-inch picture tube is being utilized in this example.) Note that the flyback pulse at the emitter of the output transistor is also rectified and filtered. Diode X508 rectifies the flyback pulse and supplies approximately +240 Vdc for the accelerating anode in the picture tube. In addition, the horizontal-output transformer supplies a reference pulse for the afc comparison circuit and a keying pulse for the keyed-agc circuit.

Horizontal-Output Configuration With Pincushion Transformer

Many modern horizontal-output circuits include a pincushion transformer for correction of curvature in raster edges. A widely used form of transistor horizontal-output circuit with a pincushion transformer is shown in Fig. 7-7A. The output transistor, Q702, functions as a switch; it is turned on to deflect the electron beam in the picture tube from the center to the right-hand side of the screen, and is then cut off for the remainder of the scanning cycle. Damper diode X701 conducts and the resulting current deflects the picture-tube beam from the left-hand side of the screen to center. Boost diode X704 also conducts on retrace pulses; it supplies about 800 Vdc to the picture-tube screen-grid circuits. In this arrangement the deflection yoke must be operated so that the horizontal-pulse voltage is balanced with respect to ground. The center of the horizontal-yoke winding is effectively at ground potential in normal operation. This balance is required to keep the pulse voltage between layers and between wires in the yoke winding within ratings, and thereby to avoid yoke breakdown.

Observe that the primary winding of the high-voltage transformer in Fig. 7-7A is split into two equal halves. That is, T710 is divided, with one winding being driven from the collector, and the other winding driven from the emitter of Q702. The collector winding is tapped at about five turns from the collector input terminal in order to provide a correct match to the yoke inductance. The supply voltage is applied to the other end of the winding; this terminal is at ac ground potential. Capacitor C721 couples the emitter pulse to the yoke via T815, and also provides pincushion correction for the horizontal sweep. Pincushion correction involves a waveshaping action whereby edge curvature is corrected by increasing the sawtooth current in the horizontal-yoke winding as the scanning beam moves down the screen from top to center. Then the sawtooth current is decreased as the scanning beam moves down the screen from center to bottom. An analogous vertical waveshaping factor is utilized to correct top and bottom pincushioning.

Note that when Q702 conducts, terminal 4 of the yoke "sees" a negative-going pulse, while at the same time terminal 10 "sees" a positive-going pulse. Although the voltage from terminal 4 to terminal 10 is nearly 1 kV at one instant, the voltage from any point in the yoke to chassis ground is less than 500 V. The emitter winding

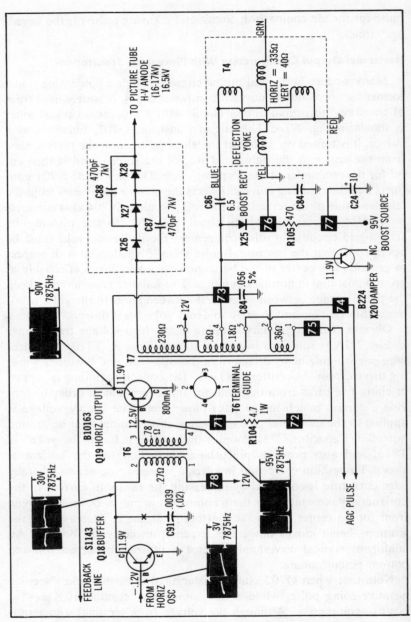

Fig. 7-8. Horizontal-output transistor arrangement with high-voltage diode-multiplier section.

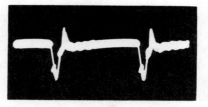

Fig. 7-9. Typical high-voltage ripple waveform.

in T710 is bifilar, and the secondary side of this winding, connected to pins 5 and 7, is used to provide a path for the centering current. As in the case of the previous transistor horizontal-output arrangement, it is essential that the driving pulse to Q702 have a very rapid rise. A slow-rise driving pulse results in lowered efficiency and in overheating of the power transistor.

Transistor Horizontal-Output Arrangement With High-Voltage Diode Stack

The transistor horizontal-output and high-voltage system shown in Fig. 7-8 is an all-solid-state design; the high-voltage rectifier section utilizes a diode voltage-multiplier network with three subsections. Note that the damper "diode" X20 is a pnp transistor with the emitter terminal "floating." As in the arrangement of Fig. 7-5 the horizontal-yoke windings are connected in parallel. The yoke, however, is returned to ground in the configuration of Fig. 7-8, whereas both ends of the yoke operate above ground in the design of Fig. 7-7A. Observe that the Q19 emitter waveform in Fig. 7-8 has a normal peak-to-peak amplitude of 90 V, whereas the supply voltage is only

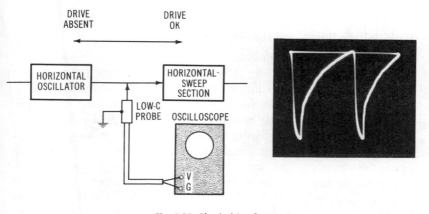

Fig. 7-10. Check drive first.

11.9 V. This voltage magnification is the result of the inductive kick-back from T7. Note also that the base drive voltage to Q19 normally has a higher peak-to-peak voltage than the emitter waveform (95 V p-p vs. 90 V p-p). Thus there is somewhat greater voltage magnification in the base circuit than in the emitter circuit. The horizontal-output transistor operates basically as a current amplifier; since its output voltage is almost equal to its input voltage, the transistor functions also as a power amplifier.

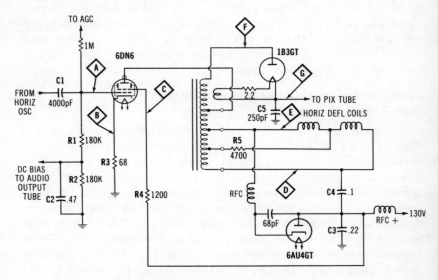

Fig. 7-11. Typical horizontal-sweep configuration.

High-Voltage Ripple and Filtering

The output from a high-voltage rectifier-filter system is never pure dc; ripple voltage (Fig. 7-9), however, can be disregarded unless its amplitude is sufficient to impair picture shading. In a simple high-voltage system the ripple frequency is 15,750 Hz, and the ripple "pulses" are comparatively narrow. Note that the ripple voltage tends to increase in amplitude as the BRIGHTNESS control is advanced; the ripple "pulse" also increases in width. If the output filter capacitor is open, the ripple may be accompanied by high-voltage "ringing." Or, if the output filter capacitor becomes leaky, the ripple is likely to include a sawtooth (exponential) component. Ripple waveforms can be checked and their peak-to-peak voltage measured with a high-

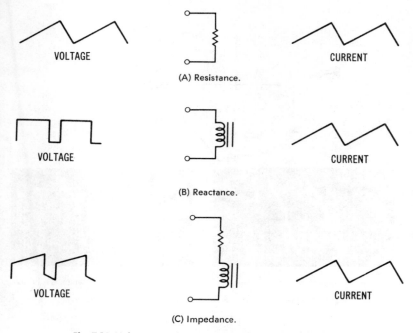

Fig. 7-12. Voltage waveforms needed for a sawtooth current.

voltage capacitance-divider probe. However, these specialized probes are seldom available except in tv labs.

TROUBLESHOOTING TUBE-TYPE HORIZONTAL-SWEEP CIRCUITRY

Television technicians are often confronted by horizontal-sweep trouble symptoms in tube-type receivers. Although this receiver section has earned a "tough-dog" reputation, troubleshooting procedures may often be facilitated by oscilloscope tests. With reference to Fig. 7-10, it is good practice to check the drive waveform to the horizontal-output tube routinely. The waveform at this point will usually sectionalize a horizontal-sweep trouble symptom. Thus, if the drive waveform is weak or absent, the trouble is almost certainly to be found in the horizontal-oscillator section. On the other hand, a normal drive waveform points to trouble within the horizontal-sweep system. One exception to this general rule is as follows: If the horizontal oscillator happens to obtain its plate-supply voltage from the

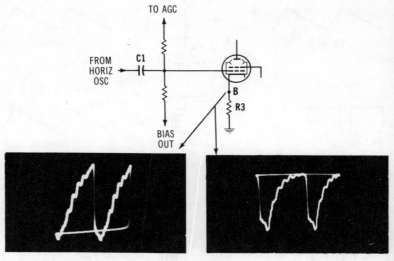

(A) Normal current waveform at point B. (B) Distorted waveform when C1 is leaky.

Fig. 7-13. Normal and abnormal waveforms at point B of Fig. 7-11.

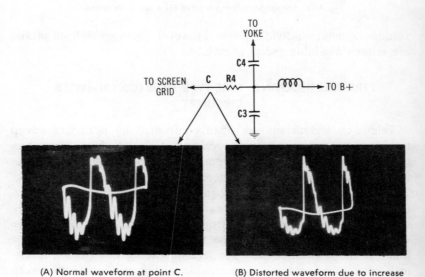

(A) Normal waveform at point C. (B) Distorted waveform due to increase in R4.

Fig. 7-14. Normal and abnormal waveforms at point C of Fig. 7-11.

B+ boost circuit, a weak drive waveform can result from sweep-circuit defects which reduce the boost voltage. In this situation, check the B+ boost voltage. If the boost voltage is subnormal, a bench power supply can be used to restore normal supply voltage while you are checking out the sweep circuitry.

Troubleshooting Procedures

Referring to Fig. 7-11, the widely used autotransformer arrangement is exemplified. The autotransformer (flyback transformer) functions to match the plate resistance of the 6DN6 output tube to the deflection-coil impedance for maximum power transfer (maximum

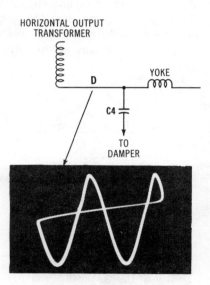

HORIZONTAL OUTPUT
TRANSFORMER

D

YOKE

C4

TO
DAMPER

Fig. 7-15. Normal waveform at point D of Fig. 7-11.

circuit efficiency). The transformer also steps up the flyback pulse voltage for processing by the high-voltage power supply. Note that the center-tapped yoke in this example has a 4700-ohm damping resistor included to minimize waveform ringing. The voltage and current waveforms in the horizontal-sweep network have different waveshapes because the load circuit is reactive (inductive). Although a complex voltage waveform will drive the same complex current waveform through a resistive load, a reactive load will characteristically modify the voltage waveshape in development of the current waveform, as depicted in Fig. 7-12. Although voltage waveforms are generally utilized in troubleshooting horizontal-sweep systems, current waveforms may be occasionally cross-checked. As an illustration, an oscilloscope check of the waveform across resistor R5 in Fig. 7-11 shows the amplitude and waveshape of the unbalanced current in the yoke circuit.

The drive waveform to the grid of the horizontal-output tube is a voltage waveform. It is checked at point A in Fig. 7-11, and it has the typical waveshape shown in Fig. 7-10. Coupling capacitor C1 has a value of 4000 pF and therefore has appreciable reactance (2500 ohms) at the 15,759-Hz scanning frequency. Whereas the horizontal oscillator supplies 90 V p-p to the coupling capacitor, only 75 V p-p are applied to the grid of the output tube. This voltage drop across the coupling capacitor is normal. However, if less than 75 V p-p were found at the grid of the output tube, the coupling capacitor would fall under suspicion—unless, of course, the horizontal oscillator was not supplying normal drive voltage. Note that

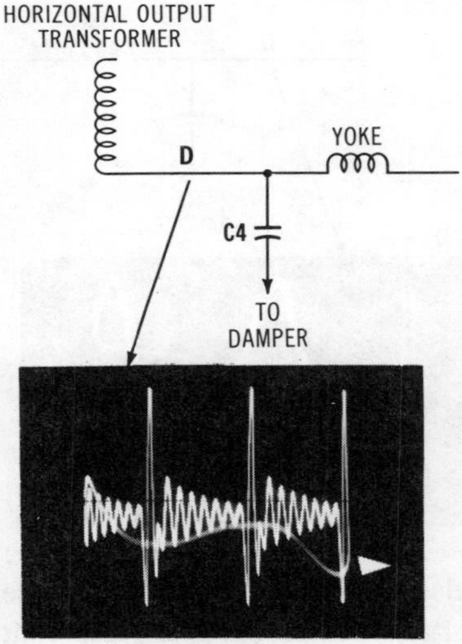

HORIZONTAL OUTPUT
TRANSFORMER

Fig. 7-16. Distorted waveform at point **D** when C4 is open.

when the capacitance of C1 is subnormal, the drive waveform at the grid not only has reduced amplitude, but also becomes distorted as a result of severe clipping of its positive peak. If C1 is completely open, no drive voltage reaches the grid of the output tube and the picture-tube screen is dark.

Low-Drive Trouble Symptoms

When the drive voltage to the horizontal-output tube is subnormal, the deflection current through the yoke is also subnormal and results in a narrow picture. Also, the high-voltage output is reduced, dimming the picture. If the BRIGHTNESS control is advanced in this situation, the picture tends to bloom. When the oscilloscope is connected across R3 in Fig. 7-11 the cathode-current waveform is displayed. The normal cathode waveform is shown in Fig. 7-13A. When C1 becomes leaky, the cathode waveform is distorted as in Fig. 7-13B. This cathode-current waveform reflects several system faults because it is the sum of plate, screen, and grid currents in the horizontal-output tube. Leakage in C1 produces a narrow picture, due to reduction of grid bias for the output tube and consequent clipping of the drive waveform.

Next, the normal screen-grid waveform for this configuration is shown in Fig. 7-14A, and a distorted waveform resulting from an increase in screen-resistor value is shown in Fig. 7-14B. When the resistance of R4 is too high, the picture shrinks horizontally. Two causes are effective in this situation: First, too high a screen resistance reduces the dc voltage at the screen grid, which limits the power output from the tube; and, second, an unbypassed screen resistor is utilized in this configuration. Thus, when the screen resistor increases in value, the signal amplitude at the screen resistor increases, although the dc voltage decreases. In turn, the screen-grid circuit operates as a triode plate-load circuit. When the load resistance increases, the output signal voltage increases. In a beam-power tube, however, the useful power is not supplied by the screen grid but by the plate. The screen-grid signal is 180° out of phase with the control-grid signal, and opposes control-grid action.

The screen grid in a beam-power tube has a lower amplification factor than the control grid, but it nevertheless has an effect on the plate current. An increase in screen-grid signal amplitude reduces the power output in the plate circuit. Technically, the screen grid has a degenerative action when the screen resistor is unbypassed. Compare this action with the cathode signal (point B in Fig. 7-11). Here the cathode signal voltage is in phase with the control-grid signal voltage. Nevertheless, a degenerative circuit action is present in the cathode circuit because a positive-going signal at the control grid increases the plate current.

Narrow Picture Analysis

With reference to Fig. 7-11, the normal waveform at point D is shown in Fig. 7-15. If C4 opens up, a very narrow picture results and the waveform at point D becomes highly distorted, as shown in

Fig. 7-16. Or, if C4 becomes leaky, the waveform amplitude is reduced although the waveshape does not change greatly, because the picture width is reduced. These examples show that for a technician who is familiar with the abnormal waveforms in a horizontal-deflection circuit, defective components can often be pinpointed. A narrow-picture symptom can be caused by more than one component defect, and the oscilloscope is often the most useful instrument to find the fault.

CHAPTER 8

Troubleshooting the Vertical-Sweep Section

A modern vertical-oscillator arrangement is shown in Fig. 8-1; the components and devices are mounted on a "vertical" module.

PRINCIPLES OF VERTICAL-SWEEP-SECTION OPERATION

In Fig. 8-1 a 34-V supply and a decoupled 30-V supply are derived from a negative-going horizontal pulse from the flyback transformer. This waveform has a normal amplitude of 240 V p-p; if the horizontal-sweep system fails, the vertical-sweep system will be automatically disabled. This flyback pulse is rectified by diode X1 and is filtered by C3, R20, and C11. Transistor Q1 is initially cut off due to a positive voltage on its base from the R1-R9 voltage divider. Capacitor C17 charges toward the 30-V level at a rate dependent upon the setting of the VERTICAL HOLD control R7-R14. The voltage on C17 is coupled to the emitter of Q1, and when it exceeds the base voltage by about 0.6 V Q1 turns on and C17 discharges. After C17 has discharged, Q1 cuts off and the cycle then repeats. The oscillator is synchronized by vertical-sync pulses from the integrator which feeds into the base of Q1. Transistors Q1, Q2, and Q3 operate in a positive-feedback loop; when Q1 turns on, Q2 and Q3 also turn on.

With reference to Fig. 8-1, capacitors C15 and C16 also charge toward the 30-V level through the VERTICAL HEIGHT control and R22.

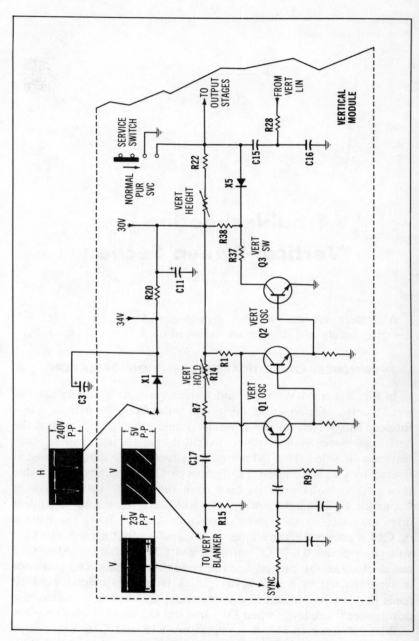

Fig. 8-1. A modular vertical-oscillator configuration.

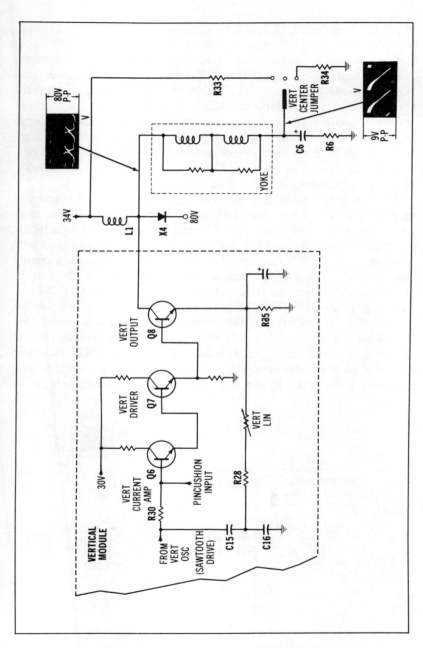

Fig. 8-2. A modular vertical-output configuration.

When Q3 turns on, C15 and C16 discharge through D5, R37, and Q3. The charging-discharging waveform at the junction of R22 and C15 is the vertical-rate sawtooth used as the oscillator output to drive the remaining vertical stages. During the vertical-retrace time (Q3 on) diode X5 clamps the junction of C15 and R22 to the voltage drop across R37 and Q3 (about 2 V). This clamping action maintains minimum bias on the output stages. A feedback waveform from the vertical-output stage is coupled to the junction of C15 and C16 through R28 for linearity correction. Note that when the service switch is in its service position, C15 and C16 discharge to ground, resulting in the vertical collapse of the scan.

The vertical-driver and output circuitry following the vertical oscillator is shown in Fig. 8-2. The output from the oscillator is coupled into the base of the vertical current amplifier, Q6, via R30. Transistors Q6, Q7, and Q8 are connected in a positive-feedback loop and about 2 V is required at the base of Q6 to bring the circuit out of cutoff. The oscillator circuit clamps the base of Q6 to this 2-V level during vertical-retrace time so that C15 and C16 do not discharge below that level. Note that pincushion correction voltage is also coupled into the vertical-sweep circuit at the base of Q6. Transistor Q7 is an emitter follower which supplies base current to the vertical-output stage, Q8. Transistor Q8 obtains collector current from the 34-V source through L1. It is a Class-A output stage whose efficiency is increased by the large series inductor, L1. When Q8 is

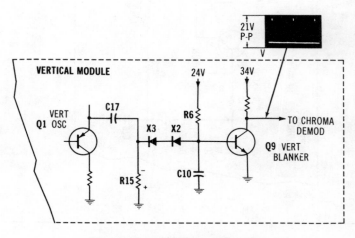

Fig. 8-3. Vertical blanking configuration.

conducting at its minimum level, maximum current is flowing through L1, the yoke, C6, and R6. This current places the scanning beam at the top of the picture-tube screen. As Q8 conducts, scanning proceeds toward the bottom of the screen. Then retrace occurs, snapping the scanning beam back to the top of the screen. Diode X4 clamps the positive retrace pulse to +80 V. The sawtooth voltage across R35 is coupled back to the junction of C15 and C16 for vertical-linearity correction through the LINEARITY control.

When the vertical centering jumper is in the center position, as shown in Fig. 8-2, no dc current flows through the yoke. However, when the jumper is moved to the R34 position, C6 and R6 are shunted by R34 and some dc current flows through the yoke and R34 to the 34-V supply. This action moves the entire raster upward slightly. Conversely, when the jumper is in the R33 position, an opposite dc current takes place through the yoke, R35, Q8, and R33. This action moves the entire raster downward slightly. Incorrect waveforms are most likely to be caused by deteriorating electrolytic capacitors. If the capacitors are not defective, however, diodes should be checked for adequate front-to-back ratio. Transistors may develop collector-junction leakage or other junction faults, and cause waveform distortion. Sometimes a short-circuited capacitor or short-circuited semiconductor device will cause an abnormal current that overheats one or more resistors, with a resulting change in resistive value.

Vertical Blanking

With reference to Fig. 8-3, a separate stage is used in this modular arrangement to provide the vertical blanking pulse to the chroma demodulator. The blanker transistor is Q9 in the diagram. Capacitor C17 has a sawtooth voltage input from the vertical oscillator, as was shown in Fig. 8-1. When Q1 turns on, C17 discharges via R15 with the result that a pulse-voltage waveform is generated as shown in Fig. 8-3. Transistor Q9 normally conducts due to the bias voltage from R6, and, in turn, the collector voltage of Q9 is low. A negative pulse is dropped across R15, and this pulse voltage forward-biases X3 and X2; thus the base end of C10 is charged negatively. As a result, Q9 cuts off and its collector voltage rises abruptly. It remains cut off until C10 discharges through R6. Then Q9 turns on and its collector voltage falls abruptly. Note that in normal operation Q9 remains cut off for the vertical-retrace interval.

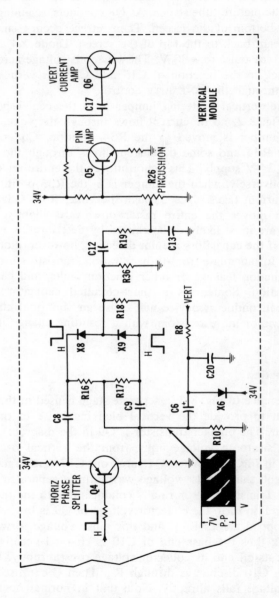

Fig. 8-4. Pincushion correction circuitry on vertical module.

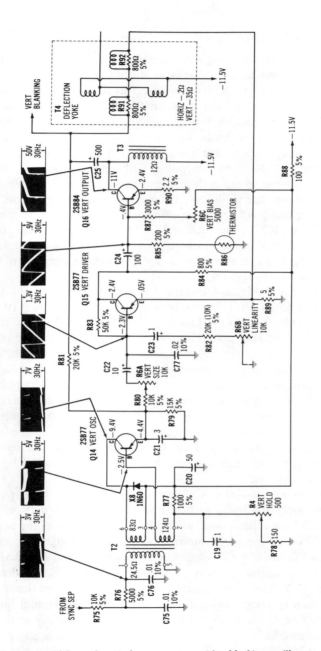

Fig. 8-5. A widely used vertical-sweep system with a blocking oscillator.

Pincushion Correction

The pincushioning effect at the sides of the raster is counteracted by means of an ac component in the dc supply to the horizontal-output stage. The pincushioning effect at the top and bottom of the raster is corrected by a circuit on the vertical module; this circuit modulates the vertical-scanning sawtooth waveform with a pincushion correction signal as shown in Fig. 8-4. A negative-going horizontal pulse is coupled into the base of Q4, a horizontal phase-splitter. Transistor Q4 develops two pulses with opposite polarity and equal amplitudes, one at its collector and the other at its emitter. These pulses are coupled to diodes X8 and X9 through capacitors C8 and C9. A vertical feedback waveform from the vertical-output circuit is coupled into R8. In turn, R8 and C20 filter the pincushion correction signal from this waveform to obtain a basic vertical-rate sawtooth. The vertical-rate sawtooth with the pincushion correction signal is shown in Fig. 8-2.

Diode X6 in Fig. 8-4 clamps the retrace portion of the feedback waveform to 34 V. This sawtooth is developed across R10 through C6 for application to the junction of R16 and R17. Since the signal at this junction is ac coupled, part of it is negative and the other part is positive. During the negative excursion of the sawtooth, X8 conducts and the negative-going horizontal pulses are added to the waveform and are developed across R36. Resistor R18 applies some positive-going horizontal pulses to the waveform at all times because more pincushion correction is desired at the top of the screen than at the bottom. The result is development of a vertical-rate sawtooth on R36, with the top half of the sawtooth modulated by negative horizontal pulses, and with the bottom half modulated by positive horizontal pulses.

Note in Fig. 8-4 that C12 passes only the horizontal pulses and thus removes the basic vertical-rate sawtooth. The network of R19 and C13 forms these horizontal pulses into parabolas for application to the base of the pincushion amplifier, Q5, through the pincushion control. Transistor Q5 amplifies these pulses and applies them to the vertical-output circuitry at the base of transistor Q6 through capacitor C17. The result is that each horizontal line on the top half of the screen is bowed upward slightly, and each horizontal line on the bottom half of the screen is bowed downward slightly to correct for the pincushioning effect. The amount of correction is adjusted by resistor R26.

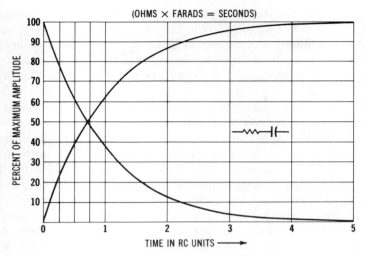

Fig. 8-6. Universal *RC* time-constant chart.

Blocking-Oscillator Arrangement

Another widely used vertical-sweep system (Fig. 8-5) employs a blocking-oscillator configuration instead of a multivibrator circuit. The windings on transformer T2 are tightly coupled and provide a large amount of feedback from the collector to the base of Q14. Transistor Q14 is reverse biased in normal operation; it operates in Class C and conducts only for brief intervals, thereby generating a pulse output waveform. During the conduction interval the base of Q14 is driven strongly by the amplified output from the collector. The base-emitter junction of Q14 rectifies these negative-going drive pulses, thereby charging C21 to a comparatively high negative voltage.

With its emitter more negative than its base, Q14 remains cut off until C21 discharges sufficiently through R79 to bring the transistor into conduction and to operate as an amplifier. Then, as Q14 comes out of cutoff, the stage provides its own input via feedback, and another surge of oscillation occurs, after which Q14 is again blocked. The exact time at which Q14 comes out of cutoff can be adjusted by varying its base-bias voltage with control R4 (VERTICAL HOLD). This control is set to a point at which the blocking rate is normally somewhat slower than 60 Hz. Negative-going sync pulses, ordinarily masked in the oscillator waveforms, are coupled to the base of Q14 via T2. *These sync pulses become visible on the oscilloscope screen if the picture is split vertically, or if the vertical oscillator is operating off frequency.* These vertical-sync pulses trigger the blocking oscillator into conduction a bit earlier than the stage would otherwise come out of cutoff, thereby achieving vertical synchronization. In case that the adjustment of the VERTICAL HOLD control is "touchy," the input sync signal to the blocking transformer should be checked first—*if the vertical-sync pulse is weak, the trouble will be found in circuitry prior to the vertical oscillator.*

Diode X8 in Fig. 8-5 operates as a "kickback limiter." It prevents the transformer from applying an excessive peak voltage to the base of Q14, which could damage the transistor. In addition, X8 has another important function; it operates as a blocking diode to prevent substantial coupling of the oscillator waveform back into the sync section. When terminal 6 of T2 is negative with respect to terminal 3, the diode effectively short-circuits the winding (there is only a fraction of a volt dropped across the diode). *In case that X8 becomes defective, operation of the sync section becomes disturbed, and sync lock may become unstable. Or, if X8 becomes open circuited, repeated failure of Q14 is probable.* These defects become apparent as an abnormal peak-to-peak voltage at terminal 1 on T2.

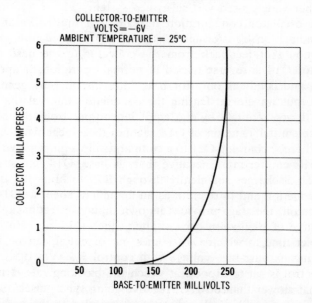

Fig. 8-7. Collector current is not proportional to base voltage in a transistor transfer characteristic.

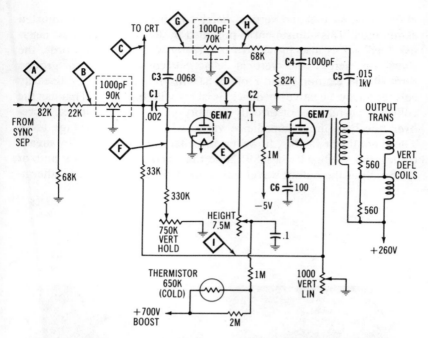

Fig. 8-8. A widely used tube-type vertical oscillator-and-sweep arrangement.

Peaker Circuit Operation

A peaked-sawtooth wave is required to obtain linear vertical deflection because the vertical-deflection coils have resistance as well as inductance. Therefore the pulse output from Q14 in Fig. 8-5 is first formed into a sawtooth wave which is subsequently peaked. The pulse is changed into a sawtooth waveform by the same circuit that provides the time constant of the oscillator (C21 and R79). Discharge of C21 through R79 develops a basic exponential waveform (curved ramp), as shown in Fig. 8-6. This waveform is linearized into a true sawtooth by negative-feedback action via R81. Circuit tolerances necessitate a manual vertical-linearity control; this manual adjustment is provided by R6B, which operates as a waveshaper in combination with C77. In consequence, the drive waveform to the base of Q15 can be made either convex or concave— between these two extremes of curvature, a linear ramp is obtained. *Inability to obtain a linear ramp indicates a component defect.*

It is also necessary to control the amplitude of base drive to Q15 in Fig. 8-5, so that the vertical height of the raster can be adjusted as required. This adjustment is provided by R6A. Observe that negative feedback is also provided for Q15 by R89. In other words, the waveform across the vertical-deflection coils returns to ground through the emitter resistor of Q15. This requirement for linearity correction by negative feedback is shown in Fig. 8-7. That is, the collector current is not directly proportional to base voltage. Moreover, the base input impedance varies with the input voltage level. These nonlinear factors, however, are effectively linearized by means of negative feedback via R89. Note that incorrect amplitude and/or distortion of the output waveform from Q15 can be caused by therm-

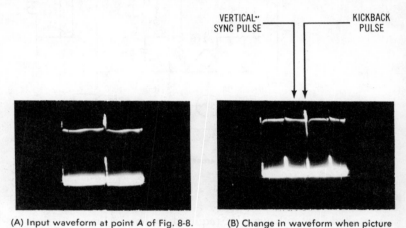

(A) Input waveform at point A of Fig. 8-8.   (B) Change in waveform when picture is split.

Fig. 8-9. Integrator input waveforms.

istor defects or by excessive leakage in the electrolytic coupling capacitor. Subnormal amplitude with distortion will result from substantial loss of capacitance in C24.

Output Stage Operation

The vertical-output stage in Fig. 8-5 employs a power transistor. Tolerances on replacement power-type transistors are comparatively wide; for this reason a bias control (R6C) is provided for maintenance adjustment of the emitter-base bias on Q16. In a power-output stage the transistor is normally operated near its maximum rated output. Since the collector current is heavy, the transistor heats up. In turn, the resistance of the semiconductor substance changes and its beta value (current amplification) increases. *Unless the compensating circuit is operating properly, the output transistor is very likely to be destroyed by thermal runaway.* The voltage waveform across the deflection coils will creep up in amplitude and then suddenly collapse as the transistor burns out. A thermistor, R86, operates to present a lower resistance in the circuit as the circuit current increases. If the current tends to increase, R86 develops a lower resistance, and the forward bias on Q16 is reduced. Reduction of forward bias decreases the base (and collector) current through Q16 and prevents thermal runaway.

Although the supply voltage to Q16 is only 11 volts, the collector-output waveform normally has a peak-to-peak voltage of 50 volts. This disparity is due to the inductive kickback from the vertical-output circuit, which also provides a peaked-sawtooth waveform.

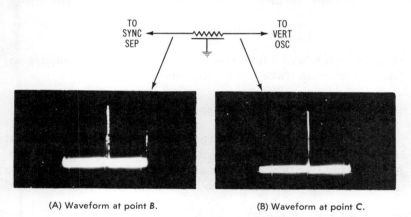

(A) Waveform at point B.   (B) Waveform at point C.

Fig. 8-10. Normal waveforms at input and output of integrator.

The negative spike on the sawtooth component is produced by inductive kickback from T3 and the vertical-deflection coils. This peaking pulse is tapped off for vertical-retrace blanking of the picture tube. Note that R90 not only provides emitter bias for Q16, but it also develops some negative feedback, inasmuch as the resistor is unbypassed. This negative feedback helps to improve deflection linearity and helps to stabilize the operating point of the transistor as its characteristics gradually change due to aging. *In case that the picture-tube screen is underscanned vertically and an oscilloscope check shows an abnormal waveform at the collector of Q16, check the yoke coupling capacitor, C25, first—it may be leaky, or have subnormal*

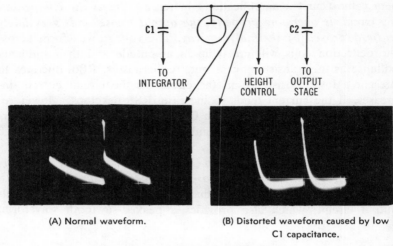

(A) Normal waveform. (B) Distorted waveform caused by low C1 capacitance.

Fig. 8-11. Waveforms at test point D.

capacitance, or have a high power factor. Then, if the coupling capacitor is normal, check out power transistor Q16.

TROUBLESHOOTING TUBE-TYPE VERTICAL-SWEEP CIRCUITRY

Although tube-type circuitry is obsolescent, tv technicians must frequently cope with older designs of vertical-sweep arrangements. A widely used vertical oscillator-and-sweep configuration is shown in Fig. 8-8. Unstable vertical-sync lock can be caused by subnormal

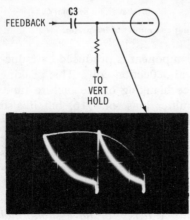

Fig. 8-12. Normal waveform at point F.

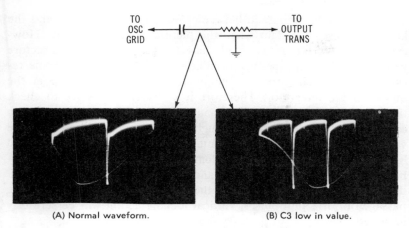

(A) Normal waveform.   (B) C3 low in value.

Fig. 8-13. Normal and abnormal waveforms at point G.

output from the sync separator; this input waveform is checked at point *A*, and normally appears as in Fig. 8-9A. It does not have the same waveshape as might be anticipated because it consists of the stripped-sync signal combined with a larger "kickback" pulse from the 6EM7. If the picture is split vertically, the kickback pulse and the vertical-sync pulse appear separately in the pattern, as shown in Fig. 8-9B. The integrator has substantial capacitance. Consequently, passage of the sync signal from test point *A* to *B* normally results in elimination of most of the horizontal-sync pulses from the waveform, as depicted in Fig. 8-10.

Since the output of the integrator is coupled to the plate of the 6EM7, the kickback pulse is comparatively large at test point *C*.

Fig. 8-14. Normal waveform at point H.

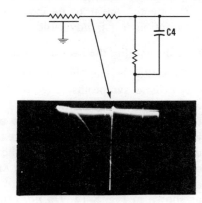

This disparity in amplitude between the vertical-sync pulse and the kickback pulse is evident when the picture is split vertically. However, it should not be assumed that this same waveform structure will be found in all tube-type oscillator-and-sweep designs. Some receivers employ circuitry that eliminates the kickback pulse at the output of the integrator. Therefore it is always necessary to check the receiver service data. Note that integrator defects, such as open capacitors, can permit feedthrough of horizontal pulses into the vertical-oscillator section. Poor interlacing is a result, accompanied by loss of picture definition. In some designs the integrator is combined with a differentiator for bandwidth control; in case that RC defects reduce the amplitude of the vertical pulse while permitting passage of horizontal pulses, vertical locking action will be seriously impaired.

Coupling-Capacitor Checks

Defective vertical oscillator-output operation is frequently caused by leakage or other faults in coupling capacitors. Referring to Fig. 8-8, C1 and C2 are ready suspects if a normal waveform (Fig. 8-11A) is not observed at test point D. These capacitors operate in a high-resistance circuit, and leakage causes off-frequency operation or, if leakage is substantial, will "kill" oscillator operation. If the coupling capacitors are in good condition, however, the HEIGHT control is another common cause of trouble symptoms. This is a very-high-resistance control, and it eventually becomes worn, with the result that it tends to become unstable and erratic. The HEIGHT control operates in combination with a thermistor, and after lengthy service the thermistor resistance may increase excessively. If these components appear to be causing the vertical trouble symptoms, a substitution test is advisable.

Troubleshooting the Feedback Loop

Capacitor leakage in the feedback loop (Fig. 8-8) can cause various trouble symptoms in vertical operation. The normal waveform at test point F is shown in Fig. 8-12. If the pattern is weak or distorted, check the waveform also at point G. The normal waveform is depicted in Fig. 8-13A; however, leakage in C3 causes a small change in the top excursion of the waveform and also speeds up the oscillator frequency—the VERTICAL HOLD control may be found out of range. A simple pulse waveform is normally found at test point H, as shown in Fig. 8-14. *Do not attempt to check waveforms between test point H and the vertical-output transformer (Fig. 8-8). The peak-to-peak voltages in this section are sufficiently high to damage a low-capacitance probe and the input circuit of an oscilloscope.* In the event that output-transformer or yoke defects are suspected, substitution tests are in order. Finally, do not overlook the possibility of leakage or loss of capacitance in cathode bypass capacitor C6—a marginal capacitor can simulate output-transformer and/or yoke trouble.

CHAPTER 9

Signal Tracing in the Sound-IF and Audio Sections

All present-day tv receivers employ a 4.5-MHz intercarrier-sound system, as exemplified in Fig. 9-1.

PRINCIPLES OF INTERCARRIER-SOUND SIGNAL PROCESSING

The sound-if amplifier has a bandwidth of 50 kHz, with a center frequency of 4.5 MHz. Sound demodulation is usually accomplished by a ratio detector having a frequency-response curve bandwidth of 50 kHz, with a center frequency of 4.5 MHz. State-of-the-art service oscilloscopes with vertical-amplifier frequency response to 4.5 MHz can be used to signal-trace the sound signal from test points A through F in Fig. 9-1. The advantage of an oscilloscope in this application is that the sound signal can be checked, not only for amplitude, but for abnormal envelope variation (interference), such as sync buzz pulses. The oscilloscope is used with a low-capacitance probe at all indicated test points. Note that audio signal waveforms are observed at test points E and F.

A popular IC sound-module assembly is shown in Fig. 9-2. It includes essentially the same circuitry as the arrangement in Fig. 9-1. The IC configuration, however, provides only two tuning slugs, compared with four in the conventional arrangement. Note that the basic oscilloscope tests are the same in either case, except that fewer test points are available in the IC design. In black-and-white receivers

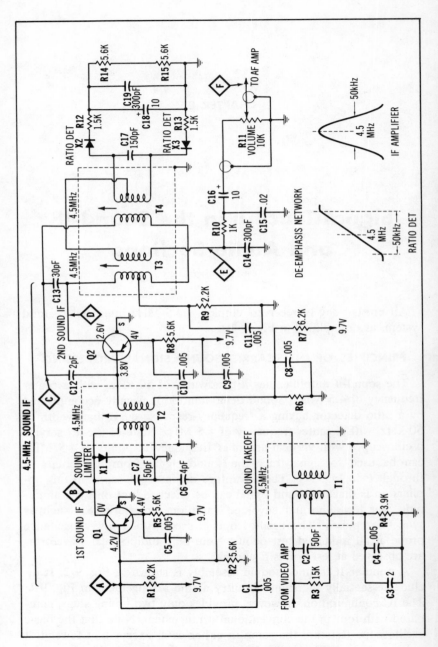

Fig. 9-1. A widely used intercarrier-if arrangement.

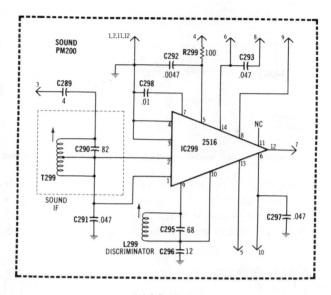

(A) Schematic.

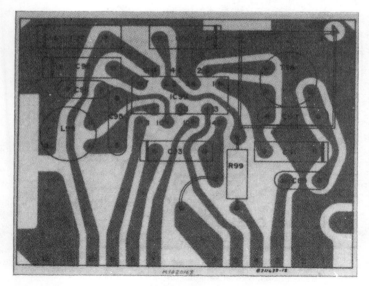

(B) Circuit board.

Fig. 9-2. A popular IC sound-module assembly.

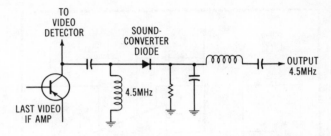

Fig. 9-3. A separate sound-converter diode is often utilized in color receivers.

the video detector does double duty as a sound converter (develops the 4.5-MHz intercarrier beat). In color receivers, however, a separate sound-converter diode is often provided, as in Fig. 9-3. This separate sound-conversion process assists in minimizing residual in-

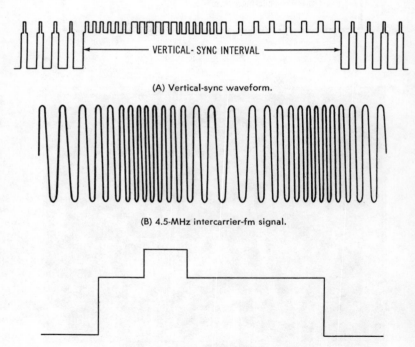

(A) Vertical-sync waveform.

(B) 4.5-MHz intercarrier-fm signal.

(C) Envelope variation displayed by scope.

Fig. 9-4. In nonlinear video-amplifier operation the vertical-sync pulse is amplitude modulated on the envelope of the 4.5-MHz intercarrier-fm sound signal.

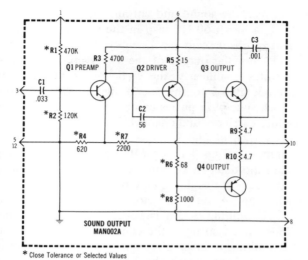

(A) Schematic.

* Close Tolerance or Selected Values

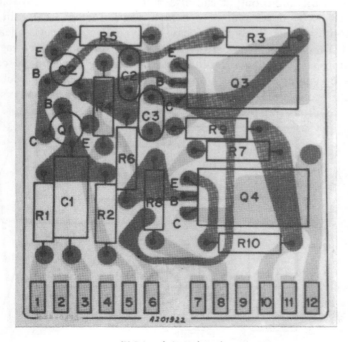

(B) Printed-circuit board.

Fig. 9-5. An extensively utilized intercarrier-sound audio module.

terference between the sound and chroma channels. Note in passing that sync buzz, a harsh 60-Hz rasp in the sound, may be a "tough-dog" problem in obsolescent black-and-white receivers that provide sound takeoff from the output of the video amplifier. If the video amplifier operates nonlinearly, the vertical-sync pulse is very likely to become modulated into the sound signal, with development of sync buzz. Such sync-buzz pulses can be traced with the oscilloscope through the intercarrier-sound channel. The development of the sync-buzz pulse is shown in Fig. 9-4.

With reference to Fig. 9-1, 4.5-MHz carrier limiting is provided by X1. If the ratio-detector response curve shows signs of distortion at higher signal levels, this diode may be open or it may have a poor front-to-back ratio. A ratio detector also provides inherent limiting action. This limiting action is dependent upon reasonably well matched diodes and also upon the condition of C18, the stabilizing capacitor. If limiting is incomplete and noise pulses appear on the audio output waveform, check the ratio-detector diodes for match. Noise in the audio output can also be caused by loss of capacitance in C18 or by development of a high power factor. Low gain in a sound-if stage is likely to be caused by collector-junction leakage in an if transistor. This defect will result in off-value dc voltages at the transistor terminals. Drift in resistor values occasionally causes trouble symptoms. For example, if R12 or R13 increases substantially in value, the S-curve will become distorted, and the audio sound output will also be distorted.

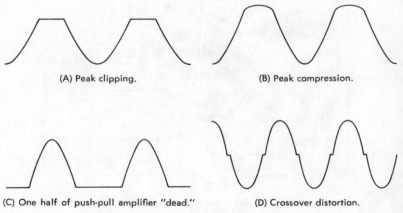

(A) Peak clipping.

(B) Peak compression.

(C) One half of push-pull amplifier "dead."

(D) Crossover distortion.

Fig. 9-6. Common forms of audio distortion.

Troubleshooting the Audio Section

A representative audio-amplifier configuration is shown in Fig. 9-5. This is a dc-coupled arrangement, in accordance with modern design. Transistor Q4 is a pnp type, whereas Q3 is an npn type; these two transistors form a push-pull complementary-symmetry stage. Transistor Q3 conducts on the positive excursion of the driver output, whereas Q4 conducts on the negative excursion of the driver output. These two transistors operate in Class AB. Audio output is taken from the common-emitter circuit. The driver transistor Q2 operates in Class A and is connected in the common-emitter mode. Similarly, Q1 operates in Class A and is connected in the common-emitter mode. Since this is a dc-coupled network, resistive values are comparatively critical—an incorrect bias value in the first stage is amplified by following stages. Negative feedback via R5, R7, R9, and R10 provides temperature stabilization for the arrangement. An audio signal can be traced through the amplifier stage-by-stage with an oscilloscope.

Common forms of audio distortion are depicted in Fig. 9-6; these distortions indicate incorrect bias conditions. Peak clipping results from a bias-voltage shift from the quiescent value specified in the receiver service data. Clipping occurs principally in high-level stages. Next, peak compression also results from a bias-voltage shift from the normal quiescent value. Compression, however, occurs chiefly in low- and medium-level stages. A "rectified" audio output waveform indicates that one of the push-pull output transistors is seriously reverse biased. Crossover distortion indicates that the bias on both output transistors is incorrect—the transistors are operating in Class B instead of Class AB. Of course, the troubleshooter also encounters combinations of these basic distortions. For example, incorrect bias voltages on the output transistors can result in a "rectified" audio output waveform with peak clipping.

Frequency distortion can also occur in the audio amplifier. Thus, if C1 in Fig. 9-5 becomes defective and loses most of its capacitance, the amplifier output will be very weak and "shrill." In other words, only the high audio frequencies are applied to the base of Q1. High-level noise interference that originates in the audio amplifier is almost always due to collector-junction leakage in a transistor. Junction leakage results also in bias shift, and the noisy output is generally accompanied by clipping or compression distortion. Leakage in C1, C2, or C3 results in bias shift and accompanying

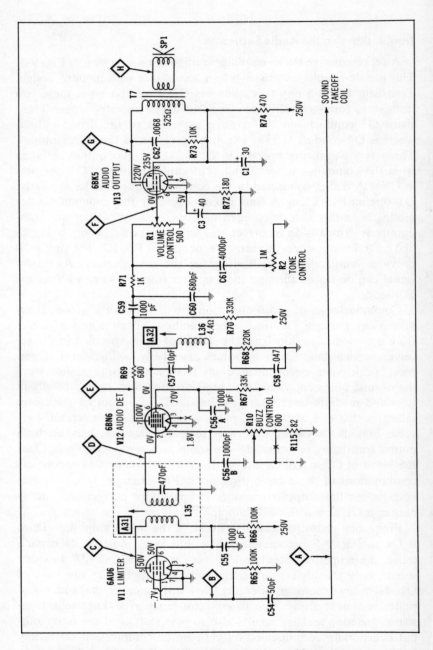

Fig. 9-7. A tube-type intercarrier-sound system used in many older designs of tv receivers.

distortion. When a transistor is replaced, it is good practice to use an exact replacement and to check the dc-voltage distribution as provided in the receiver service data. Although the oscilloscope is a very useful instrument for tracking down the source of substantial audio distortion, it cannot directly indicate small percentages of distortion. For example, it is impractical to "see" 1-percent distortion in a sine-wave pattern; unless the oscilloscope operator is experienced, he or she will have difficulty in "seeing" 3-percent distortion. This is not a matter for concern, however, inasmuch as tv sound systems are not designed for high-fidelity reproduction.

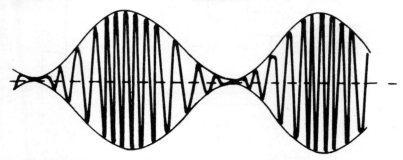

Fig. 9-8. A 100-percent amplitude-modulated output signal from am generator with incidental frequency modulation.

TROUBLESHOOTING TUBE-TYPE INTERCARRIER-SOUND SYSTEMS

Television technicians routinely troubleshoot and repair various tube-type intercarrier-sound systems although these are obsolescent. A configuration that has been used in many older designs of receivers is shown in Fig. 9-7. It contains a pentode limiter-amplifier, a quadrature (gated-beam) detector, and a beam-power audio-output tube. If an amplitude-modulated 4.5-MHz signal from a conventional signal generator is applied through a small blocking capacitor to the sound takeoff point, an oscilloscope with 4.5-MHz response can be used to trace the signal through the circuitry. Although this might seem to be an impossible procedure, inasmuch as a limiter removes the amplitude modulation from a carrier, it is a usable method for two reasons. First, service-type am generators have more or less incidental-fm output, particularly when operated at high-percentage modulation (Fig. 9-8). Second, a single limiter stage is not completely effective in rejecting amplitude modulation, and a certain percentage of the am envelope will pass. If the test signal is slightly detuned, the incidental-fm component becomes more prominent in the intercarrier-if circuit due to slope detection.

Referring to Fig. 9-7, if the test signal is found at test points *A* and *B,* the sound takeoff circuit is known to be workable (it can be peaked at 4.5 MHz for maximum output). At point *C* an amplified 4.5-MHz signal is normally displayed, but with much less apparent

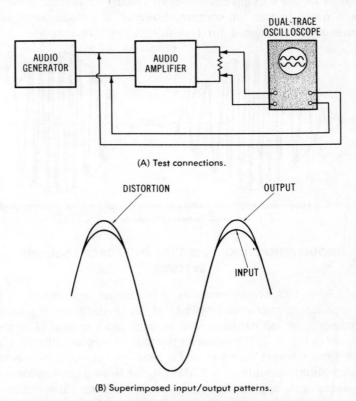

(A) Test connections.

(B) Superimposed input/output patterns.

Fig. 9-9. Checking for audio distortion with dual-trace oscilloscope.

amplitude modulation than at point *B*. Note that both the signal grid and the quadrature grid (pin 6) in the gated-beam detector are biased to provide additional limiting action. However, the test signal normally passes through the detector to point *E* because of the incidental-fm component in the am signal. In turn, the audio signal at point *E* can be traced through the audio-amplifier section at points

F, G, and *H.* Electrolytic capacitors are ready suspects in case of weak output. Distorted output is often caused by coupling-capacitor leakage.

Note the BUZZ control, R10, in Fig. 9-7. This potentiometer provides adjustment of the limiting level in the gated-beam detector tube, thereby permitting maximum rejection of 60-Hz sync buzz. This adjustment is often found to be somewhat critical. In some troubleshooting situations, adjustment of the BUZZ control does not suffice to reduce the sync-buzz level adequately. In such a case an oscilloscope check at the control grid of the gated-beam detector tube will usually show that there is an abnormal "downward" modulation of the 4.5-MHz carrier by the vertical-sync pulse. This difficulty has its source in the video-amplifier circuitry in which the sound takeoff coil is connected. It will be found that the video-amplifier stage is incorrectly biased, or that it is being overdriven, so that the vertical-sync pulse is being extensively modulated into the 4.5-MHz carrier. Therefore, video-amplifier operation must be linearized in order to correct the sync-buzz trouble symptom.

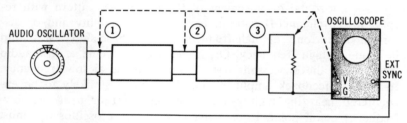

Fig. 9-10. Relative-phase checks are made with the aid of the external-sync function of the oscilloscope.

Oscilloscope Check for Linear Operation

Linear operation of an amplifier requires that the output waveform be the same as the input waveform; the amplifier output at any point on its path of operation should be directly proportional to the amplifier input. A dual-trace oscilloscope provides a critical quick-check of amplifier linearity, as shown in Fig. 9-9. This is a linearity test for an audio amplifier; a video-amplifier test can be made in the same manner. An audio generator applies a sine-wave signal voltage to the input of the amplifier. A power resistor of suitable ohmic value is connected across the output terminals of the amplifier. One vertical channel of a dual-trace oscilloscope is connected across the input terminals of the amplifier, while the other vertical channel of the oscilloscope is connected across the output terminals of the amplifier. The channel gains are equalized, and the input/output waveforms are superimposed on the oscilloscope screen. If the two waveforms coincide precisely, the amplifier is operating linearly. On the other hand, if the two waveforms do not coincide, as exemplified in Fig. 9-9B, the amplifier is distorting and is not operating linearly.

Phase-Shift Tests

Phase relations are sometimes significant in audio troubleshooting procedures. Distortion and instability in negative-feedback arrangements often result from abnormal phase shift due to defective components. Progressive phase shift through a system can be easily checked with a dual-trace oscilloscope. Thus the input signal to a feedback loop, for example, can be fed into the A channel of the oscilloscope, and the output signal from the feedback loop can be fed into the B channel. Any phase shift that may be occurring shows up as a horizontal displacement of the channel-A pattern with respect to the channel-B pattern. Note that if instability and/or distortion occurs only at high frequencies, the phase-shift test should be made at high frequency. Or, if the trouble occurs at low audio frequencies, the phase-shift test should be made at low frequency. In negative-feedback amplifiers, instability is most likely to be encountered near the high-frequency cutoff point or near the low-frequency cutoff point. In case that a single-trace oscilloscope must be used in phase-shift tests, the external-sync function should be used, as shown in Fig. 9-10. When this method is employed, the pattern at (1) is taken as the phase reference; its phase is compared with that of the pattern at (2), and of the pattern at (3).

CHAPTER 10

Digital-Logic Troubleshooting
With the Oscilloscope

Digital-logic equipment is primarily a network of *gates* and *flip-flops*. We will see that a flip-flop is an arrangement of cross-connected gates. Digital-logic equipment operation is *synchronized* by pulses (clock pulses) produced by a clock generator. Clock generators operate at various speeds, up to many megahertz. Oscilloscopes used in troubleshooting digital-logic equipment may therefore require unusually high vertical-amplifier frequency response. For example, a typical gate can be operated by 30-MHz square waves. When several gates are connected into a network, however, the system operating speed is necessarily less than that of a single gate.

DIGITAL-LOGIC WAVEFORM RELATIONSHIPS

The four most basic forms of gates are the AND gate, the NAND gates, the OR gate, and the NOR gate, as shown in Fig. 10-1. Each gate has two inputs in these examples; later we will encounter gates that have three, four, and more inputs. The number of inputs to a gate does not change its basic mode of response, however. Observe in Fig. 10-1A that a square-wave input and a rectangular-wave input are applied to the AND gate. In turn, two square waves normally appear at the output. In other words, *an AND gate produces an output only when both of its inputs are driven to logic high.* During the time that either input of the AND gate is zero (logic low), there is zero output from the AND gate. However, during the time that both inputs of the AND gate are at logic high, there is a logic-high output from the AND gate. Note next that a NAND gate (Fig. 10-1C) operates in the same manner as an AND gate, except that its output is inverted. That is, the output from a NAND gate is a logic high during the time that both of its inputs are at logic low. When both inputs of the NAND gate are driven to logic high, however, the output of the NAND gate will go to logic low. Suppose that a NAND gate is provided with three

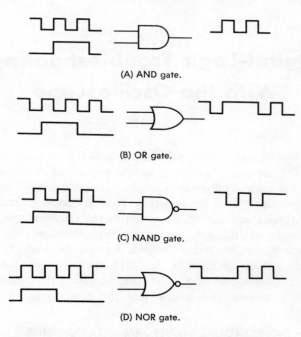

(A) AND gate.

(B) OR gate.

(C) NAND gate.

(D) NOR gate.

Fig. 10-1. Signal responses of AND, OR, NAND, and NOR gates.

inputs; then the NAND gate output is at logic high until all three inputs are driven to logic high. With all three inputs simultaneously at logic high, the NAND gate output will go logic-low.

Next, observe the operation of the OR gate in Fig. 10-1B. Note that *an OR gate produces an output if either of its inputs is driven to logic high; the OR gate also produces an output if both of its inputs are driven to logic high.* If an OR gate is provided with four inputs, it will produce an output when one or more of the inputs are driven to logic

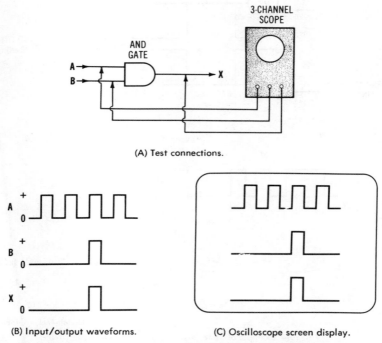

(A) Test connections.

(B) Input/output waveforms.

(C) Oscilloscope screen display.

Fig. 10-2. Check of AND gate operation with a three-channel oscilloscope.

high. Note next that a NOR gate operates in the same manner as an OR gate, except that its output is inverted. In other words, the output terminal of a NOR gate is at logic high as long as both of its inputs are at logic low. However, if one or both inputs of the NOR gate are

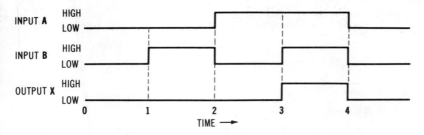

Fig. 10-3. A timing diagram for an AND gate.

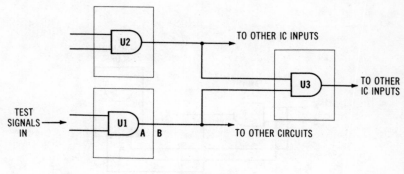

(A) Circuit.

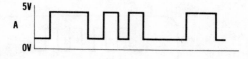

1.4V to 1.5V = "Bad Level" and is Interpreted by TTL and DLT Inputs as a High State

(B) Signals at points A and B.

Courtesy Hewlett-Packard, Inc.

Fig. 10-4. Example of "bad level" output due to an open circuit.

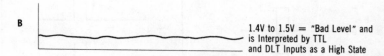

Fig. 10-5. Logic-high and logic-low thresholds in TTL and DLT circuitry.

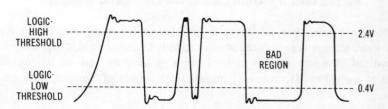

(A) Good signal.

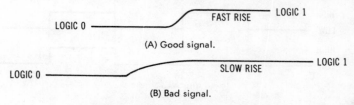

(B) Bad signal.

Fig. 10-6. Excessive rise time is a fault condition.

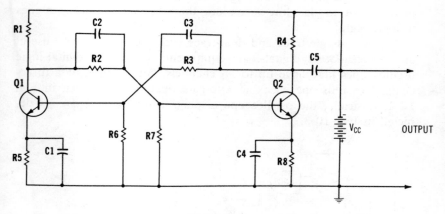

(A) Circuit.

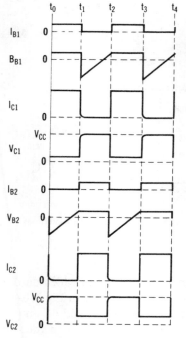

Voltage Waveforms can be Displayed Directly; Current Waveforms Cannot be Displayed Unless Indirect Test Techniques are Use.

(B) Voltage waveforms can be displayed directly; current waveforms cannot be displayed unless indirect test techniques are used.

Fig. 10-7. Basic clock oscillator and normal waveforms.

driven to logic high, its output goes to logic low. Then, when both of the inputs go to logic low once more, the NOR gate output again goes to logic high.

Although single-trace and dual-trace oscilloscopes are very useful in troubleshooting digital-logic equipment, specialized digital-logic oscilloscopes may have up to 16 channels. An example of a three-channel oscilloscope checking AND-gate operation is shown in Fig. 10-2. Basically, the oscilloscope displays a *timing diagram,* as depicted in Fig. 10-3.

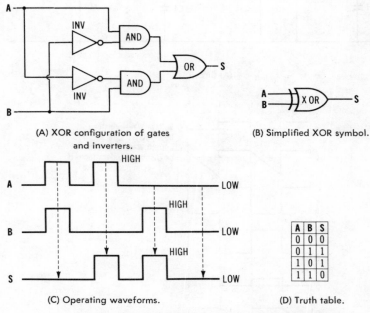

(A) XOR configuration of gates and inverters.

(B) Simplified XOR symbol.

(C) Operating waveforms.

(D) Truth table.

A	B	S
0	0	0
0	1	1
1	0	1
1	1	0

Fig. 10-8. Exclusive-OR (XOR) gate and operating waveforms.

Fault Patterns

Consider a unit of digital-logic equipment in which an open circuit occurs at the output of an AND gate, as shown in Fig. 10-4. Test signals are applied at the inputs of AND gate U1, and an output signal is displayed at the output of the gate. However, when the oscilloscope probe is moved ahead in the same circuit, there is only a dc "bad level" output. In TTL and DTL logic circuitry the range from 1.4 to 1.5 volts indicates a fault (in this example, an open circuit); note

234

that a "bad level" will be interpreted by following circuits as a logic-high level. A logic-high level is at least 2.4 V, and a logic-low level is 0.4 V or less, as depicted in Fig. 10-5. The output waveform from a gate must rise and fall within a time sufficiently short that subsequent gates in the system can be operated properly. Thus excessive rise time is another basic fault condition, as exemplified in Fig. 10-6. Note that the rise time of a digital signal will always be slowed more or less by lines (interconnects) that are installed for operation of peripheral memories or other logic units. The clock rate may therefore need to be decreased if interconnects impose excessive rise time.

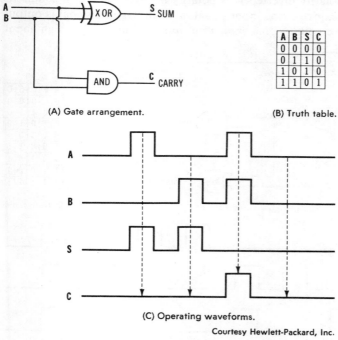

A	B	S	C
0	0	0	0
0	1	1	0
1	0	1	0
1	1	0	1

(A) Gate arrangement.

(B) Truth table.

(C) Operating waveforms.

Courtesy Hewlett-Packard, Inc.

Fig. 10-9. Half adder.

OPERATING WAVEFORMS IN SIMPLE DIGITAL NETWORKS

A basic clock oscillator (multivibrator) is shown in Fig. 10-7, with normal operating waveforms. In a typical system, V_{CC} may be 6 V. Faulty operation in this type of circuitry can result either from defective capacitors or failing transistors. Note that power failure occurs on occasion. This may be caused by trouble in the power supply itself, in the power distribution cabling, or in the power de-coupling circuitry included on each circuit board. Another common cause of trouble symptoms is a connecting wire that has become open or shorted to ground. Transistors tend to fail catastrophically. Short-circuited elements may overheat and damage resistors in the associated circuit.

Half Adders and Full Adders

Observe next the exclusive OR (XOR) gate shown in Fig. 10-8. An XOR gate can be formed from two AND gates, an OR gate, and two inverters (polarity inverters). A polarity inverter, such as a common-emitter stage, inverts the input signal; a positive-going signal is output as a negative-going signal, and vice versa. Thus a logic-high input

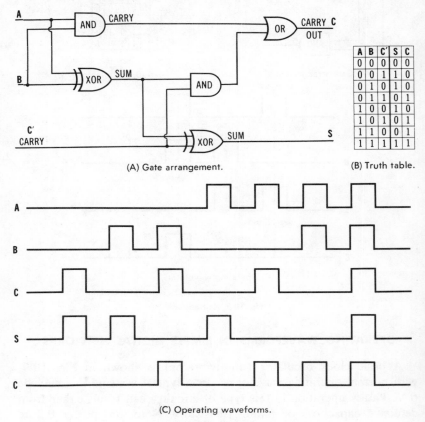

(A) Gate arrangement.

A	B	C'	S	C
0	0	0	0	0
0	0	1	1	0
0	1	0	1	0
0	1	1	0	1
1	0	0	1	0
1	0	1	0	1
1	1	0	0	1
1	1	1	1	1

(B) Truth table.

(C) Operating waveforms.

Fig. 10-10. Full adder.

signal is output as a logic-low signal by an inverter. An XOR gate functions to produce a logic-high output when one of its inputs is at logic high and its other input is at logic low. On the other hand, the output of an XOR gate remains at logic low when both of its inputs are at logic low, or when both of its inputs are at logic high. In other words, *an XOR gate produces an output only when its inputs are at opposite logic levels; an XOR gate produces zero output when its inputs are at the same logic level.* An XOR gate responds to input pulses as shown in Fig. 10-8C. An XOR gate with an AND gate is called a *half adder* because its output is the sum of 0 and 1, or of 1 and 0, or of 1 and 1 (Fig. 10-9).

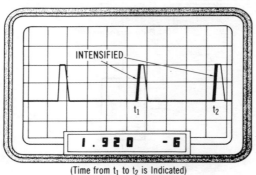

(Time from t_1 to t_2 is Indicated)

Fig. 10-11. An "intelligent oscilloscope" can indicate the elapsed time between the two operator-positioned intensity markers in a digital-logic waveform.

A *full adder* consists of two half-adders and an OR gate, as shown in Fig. 10-10. A full adder provides for a carry in, C', as well as for a carry-out, C, terminal. Provision of a carry-in line permits connection of two full adders in series, for the simultaneous (parallel) addition of large binary numbers.

Measuring Time Intervals

In addition to checking the timing diagram for a unit of digital-logic equipment, the pulse widths and the elapsed time between pulses can be measured wtih a triggered-sweep oscilloscope that has a calibrated time base. Note that if an "intelligent" oscilloscope is employed, the operator can place intensity markers at the start and stop intervals to be timed (t_1 and t_2 in Fig. 1-11). Then the LED readout in this example automatically and continuously indicates the time between the two markers (1.92×10^{-6} second). This feature provides

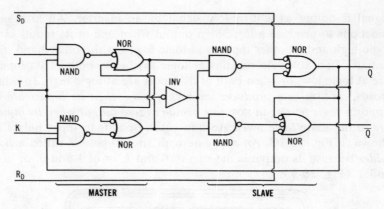

(A) Gate arrangement.

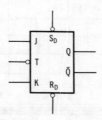

(B) Logic symbol.

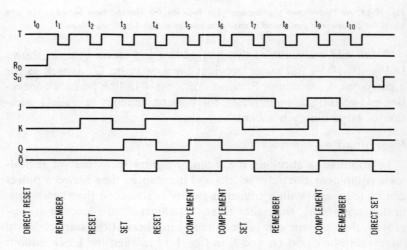

(C) Waveform relations (timing diagram).

Fig. 10-12. Operation of J-K flip-flop.

operator convenience, inasmuch as the calculation of time is performed automatically.

Binary Counters

Binary counters are widely used in digital equipment. Many counters employ J-K flip-flops, as shown in Fig. 10-12. A J-K flip-flop has two conditioning inputs (J and K) and one clock (T) input. If both conditioning inputs are disabled prior to a clock pulse, the flip-flop does not change condition when a clock pulse occurs. If the J input is enabled, and the K input is disabled, the flip-flop will assume the 1 condition (Q output at logic high; $\overline{Q}$ output at logic low) upon arrival of a clock pulse. Note that $\overline{Q}$ denotes NOT Q, or the inverse (complement) of Q. Next, if the K input is enabled, and the J input is disabled, the flip-flop will assume the 0 condition when a clock pulse arrives. If both the J and K inputs are enabled prior to the arrival of a clock pulse, the flip-flop will complement, or assume the opposite state, when the clock pulse occurs. In other words, at the application of a clock pulse, a 1 on the J input sets the flip-flop to the 1 or "on" state; a 1 on the K input resets the flip-flop to its 0 or "off" state. If a 1 is simultaneously applied to both inputs, the flip-flop will change state, regardless of its previous state.

Next, observe the 4-bit binary counter arrangement shown in Fig. 10-13. A bit denotes a *bi*nary digi*t,* such as 1 or 0. Examples of 4-bit binary numbers are:

0000 = zero
0001 = one
0010 = two
0011 = three
0100 = four
0101 = five
0110 = six
0111 = seven
1000 = eight
1001 = nine
1010 = ten
1011 = eleven
1100 = twelve
1101 = thirteen
1110 = fourteen
1111 = fifteen

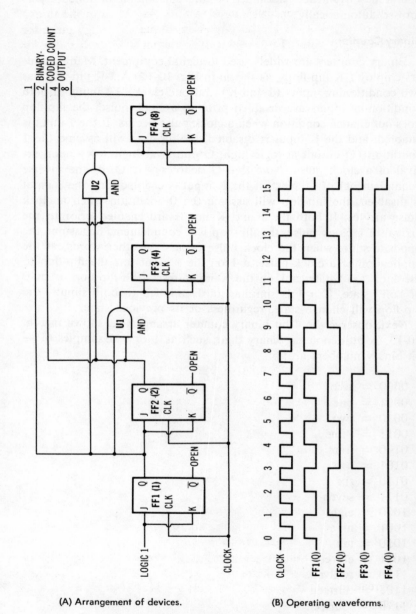

(A) Arrangement of devices.

(B) Operating waveforms.

Fig. 10-13. A 4-bit binary counter circuit.

The counter depicted in Fig. 10-13 counts from zero (0000) to fifteen (1111), and then resets itself to zero. As shown in the timing diagram of Fig. 10-13B FF1 produces an output logic-high pulse for each two input pulses; FF2 produces an output logic-high pulse for every four input pulses; FF3 produces an output logic-high pulse for every eight input pulses; FF4 produces an output logic-high pulse for every sixteen input pulses. Thus this is also called a *divide-by-16* configuration. Observe that the J and K inputs are tied together on each flip-flop; this is called the *toggle* mode of operation. In other words, FF1 will change state on every clock pulse, provided only that the logic-1 input is held at logic high. Of course, if the logic-1 input is driven to logic low, the counter stops operation. After the count reaches fifteen (1111), the next clock pulse will reset the chain to 0000 because each flip-flop must necessarily change state from 1 to 0 at this time. The AND gates are included in this counter design simply to obtain faster response (minimum propagation delay).

Another term for the 4-bit binary counter arrangement in Fig. 10-13 is *synchronous binary counter*. If the AND gates are omitted, the flip-flop chain is then called a *ripple-carry counter*. A ripple-carry counter requires a certain amount of time to change state. This state-change time involves more or less propagation delay. As an illustration, suppose that all four flip-flops in Fig. 10-13A are in their 1 state. On the next clock pulse, FF1 changes to its 0 state; after FF1 changes to its 0 state, FF2 changes to its 0 state; after FF2 changes to its 0 state, FF3 changes to its 0 state, and so on. This sequence is called *ripple-carry action*. Observe that when the AND gates are included, counter action is speeded up because all flip-flops are then triggered simultaneously from the input. Thus the chain changes state from 1111 to 0000 simultaneously, instead of sequentially, when the synchronous configuration is used. As a result, the clock can be operated at a considerably higher frequency.

Digital Pulse Waveshapes

In theory, digital pulses have square corners; the leading edge ideally has zero rise time, and the trailing edge has zero fall time. In practice, however, substantial departures from the ideal often occur. In one type of digital equipment the pulses have an appearance as shown in Fig. 10-14. The tops of the pulses are rounded, the rise time is greater than zero, and the fall time is longer than the rise time. Thus the pulses are somewhat spike shaped instead of being precisely rectangular. However, this is not a matter for practical concern, as

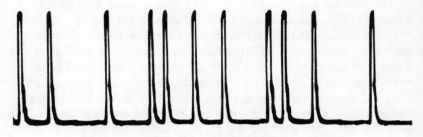

Fig. 10-14. Digital pulses may not have ideal shape in practice.

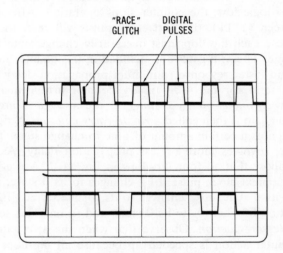

(A) Race glitch (see Fig. 10-16).

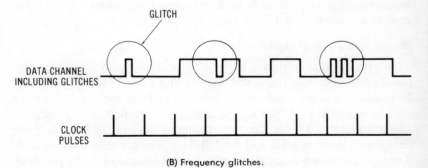

(B) Frequency glitches.

Fig. 10-15. Examples of glitches in digital data signals.

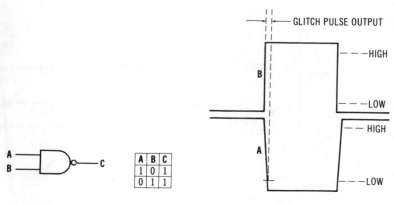

(A) NAND gate and truth table.

A	B	C
1	0	1
0	1	1

(B) How a race glitch happens.

Fig. 10-16. A race condition occurs if B goes to logic high before A attains the logic-low level.

long as the distortion is not so serious that the gates malfunction. Tolerable waveshape distortion is chiefly a matter for experienced judgment; it is difficult to set down hard-and-fast rules in this regard.

Propagation delay was noted previously. While not a form of pulse distortion *per se,* excessive propagation delay in a digital system inevitably leads to unreliable operation. Therefore in troubleshooting a digital system with the oscilloscope the technician should be alert for abnormal propagation delays, as well as for excessive pulseshape distortion.

Excessive propagation delay and waveshape distortion can give rise to *glitches* in otherwise well-designed digital equipment. Glitches are spurious pulses produced by marginal gate operation. Examples of glitches are given in Fig. 10-15. It is evident that glitches will cause equipment malfunction, inasmuch as a gate or a flip-flop cannot "tell the difference" between a glitch and a data pulse or waveform. Glitches can also occur because of power-supply switching transients. Sometimes glitches are induced in digital equipment by strong external fields, as in industrial installations. In any case the oscilloscope is the most useful troubleshooting instrument for analyzing and tracking down the source of glitches.

Note also that glitches are occasionally very narrow and may have very fast rise and fall. In these instances a high-performance oscilloscope is required to display a visible glitch pattern. Considerably

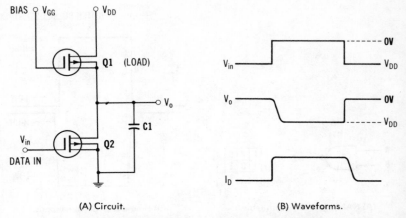

(A) Circuit. (B) Waveforms.

Fig. 10-17. Normal rounding of digital pulses by an MOS ratio-inverter device.

higher sweep speeds may be required than for displaying the normal operating waveforms in the digital equipment. Troubleshooting procedures are sometimes impeded by transient occurrence of a glitch. If the technician knows that a trouble symptom is being caused by a glitch, but the glitch occurs randomly at widely separated intervals, it may be a time-consuming task to identify the trouble area and to close in on the defective device or component. In such a case it is often helpful to use a storage-type oscilloscope as a monitor. The technician can then look at the screen from time to time, to determine whether a glitch has been displayed. It relieves the oscilloscope operator from the necessity of closely and continuously watching the screen to "catch the glitch on the fly." In practice, glitches as narrow as 25 ns are encountered, and only the most sophisticated oscilloscopes are capable of "catching" and displaying such narrow pulses.

A *race* condition is present in a digital circuit when two or more pulse inputs to the circuit arrive simultaneously. It is apparent that *if the* order *in which the two pulses are applied to a device determines its output state, a critical race condition exists.* Refer to Fig. 10-16A; this is a two-input NAND gate which can malfunction if input B goes to logic-high *before* input A attains the logic-low level. In this situation A and B are both driven to logic high momentarily, with the result that a spurious pulse is produced in the output, C. Of course, the spurious pulse ends as soon as input A reaches the logic-low threshold. This negative pulse is a typical glitch. In digital-equipment design, propagation delays inherent in the A and B input circuits may be "tailored" to avoid a race condition. Avoidance of glitches then depends on permissible waveform tolerances. Thus, in the event that the pulse input to A develops excessive fall time, a race condition will occur, with development of glitch output from the NAND gate.

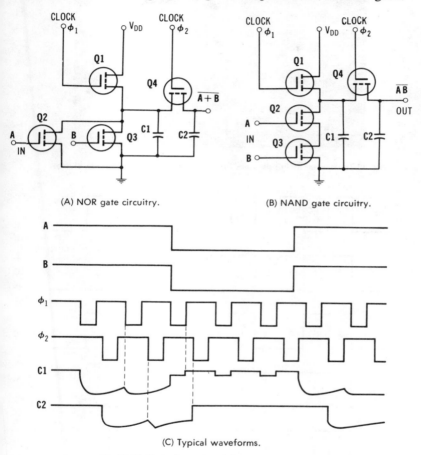

(A) NOR gate circuitry.

(B) NAND gate circuitry.

(C) Typical waveforms.

Fig. 10-18. Basic dynamic ratioless two-phase gates.

In Fig. 10-17 the rounding of the output pulse from the MOS ratio-inverter device is normal and should not be a matter for concern to the troubleshooter. Of course, excessive corner rounding accompanied by abnormally long rise time can cause equipment malfunction. Permissible tolerances on waveshape are largely learned by

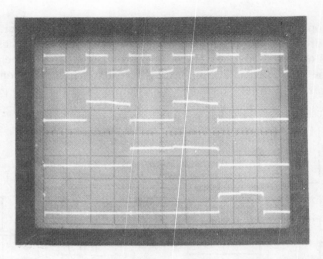

Fig. 10-19. Normal timing diagram for the Q outputs of a TTL decade divider.

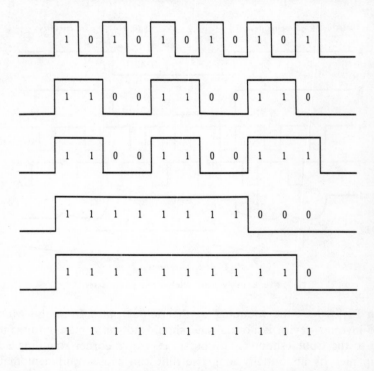

Fig. 10-20. A representative correlation of timing-diagram and data-domain displays.

experience. For example, the distorted waveforms across C1 and C2 in Fig. 10-18 could well be regarded with alarm by the apprentice technician. However, these waveforms are entirely normal for these MOS devices, as may be determined by consulting the manufacturer's technical data. Normal waveforms are less than ideal in TTL circuitry; for example, the four-channel display in Fig. 10-19 for the Q outputs of a TTL decade divider shows noticeable departures from the ideal waveshapes. Thus the clock pulses exhibit some tilt along their bottom excursions; the next lower waveform has clearly visible tilt along its top excursions; all of the pulses show some evidence of overshoot or undershoot.

DATA-DOMAIN DISPLAYS

Some digital-troubleshooting oscilloscopes provide a choice of timing-diagram or data-domain displays, as shown in the correlation example of Fig. 10-20. Thus the troubleshooter may switch his or her six-channel oscilloscope either to display timing diagrams or to display data fields. In preliminary analysis of equipment malfunction it is often quite helpful to observe the data-domain display of binary digits. This display can be directly compared with truth tables, and requires no "interpretation" from wavetrains into digital "words." In turn, the channel that exhibits a malfunction is directly identifiable. On the other hand, analysis of the malfunction can generally be made to better advantage on the basis of a timing-diagram waveform display. For example, if the malfunction happens to be caused by a glitch, the spurious pulse can be tracked down more confidently on the basis of a conventional waveform display.

Index

A

Ac
  input coupling, 14
  waveforms with dc components,
    65-66
Action of triggered-sweep controls,
    69-80
Adders, half and full, 236-237
Afc and horizontal-oscillator section,
    troubleshooting, 165-180
Amplitude modulation, checking, 35
AND gate, 229-230
Astigmatism control, 45
Audio and sound-if sections, signal
    tracing in, 217-228
Average voltage, sine wave, 13

B

Balanced afc network, 165
Base line offset, distortion factor, 19
Basic
  oscilloscope with free-running
    sweeps, 41-46
    centering-control adjustment,
      43-44
    focus-control adjustment, 44-46
    intensity-control adjustment,
      41-43

Basic—cont
  waveforms, 13-35
    cornering of, 30
    growth and decay, 22-26
    hidden distortion, 19-21
    Lissajous family, 32-35
    modified sine, 29-30
    pulse polarities, 30-32
    steady state vs. transient, 26-29
Binary counters, 239-241
Bit, 239
Blocking oscillator, vertical-sweep
    section, 207-208
Burst waveforms, 28-29

C

Calibration, vertical circuit, 62
Capacitive voltage division, probe, 86
Capacitance-divider probe, high volt-
    age, 95-99
Centering-control adjustment with
    free-running sweep, 43-44
Circuit ghosts, 122
Clipping
  and compression, waveform, 131
  in vertical amplifier, 56
Clock
  digital, 11

249

Clock—cont
  waveform distortion factors, 18-19
    base line offset, 19
    fall time, 19
    overshoot, 18, 30
    preshoot, 18
    ringing, 19, 30
    rise time, 18
    rounding, 18, 30
    sag, 19
    settling time, 19
Coarse and fine deflection controls, 47
Color-sync troubleshooting with
    scope, 150-159
Composite video, sync lock, 50
Compression and clipping, waveform,
    131
Construction and operation of demod-
    ulator probes, 88-95
  double-ended probe, 94-95
  low-impedance probe, 92-94
  medium-impedance probe, 92
  shunt-detector probe with RC filter,
    90-91
Control functions and terminal facili-
    ties, 46-58
  coarse and fine deflection controls,
    47
  display of 60-Hz test voltage, 46-47
  gain-control ranges and settings,
    55-57
  intensity-control settings vs. pattern
    size, 53-54
  sync-control adjustment, 47-50
Cornering of waveform, 30
Counters, binary, 239-241
Coupling-capacitor checks, vertical-
    sweep section, 214
Cycle, duty, 16
Cycles per second, see Hertz

D

Damped sine wave, 29
Data-domain displays, 247
Dc voltage components and measure-
    ments, 63-65
Delay, propagation, 243
Delay control, 74

Demodulator probes, construction and
    operation, 81-95
  double-ended, 94-95
  low-impedance, 92-94
  medium-impedance, 92
  shunt-detector with RC filter, 90-91
Detectors, traveling, 92
Digital
  -logic troubleshooting, 229-247
    data-domain displays, 247
    operating waveforms, 235-247
    waveform relationships, 229-235
    pulse waveshapes, 241-247
Diode stack, high-voltage, 191-192
Direct-cable input, 83
Displays
  data-domain, 247
  of 60-Hz test voltage, 46-47
Distortion
  factors, clock waveform, 18-19
    base line offset, 19
    fall time, 19
    overshoot, 18, 30
    preshoot, 18
    ringing, 19, 30
    rise time, 18
    rounding, 18, 30
    sag, 19
    settling time, 19
  hidden, 19-21
Double-ended demodulator probe,
    94-95
Dual
  -mode oscilloscopes, 69
  -time-constant sync-clipper action,
    155-159
  -trace display modes, 79-80
Duty cycle, 16

E

Exclusive-OR gate, 236-237
Exponential waveform, 22
External Sync terminal, 67

F

Factors, distortion, clock waveform,
    18-19
Fault patterns, digital-logic, 234-235

Fall
  pulse, 18
  time, 18
Feedback loop, vertical-sweep section,
  214-215
Fine and coarse deflection controls, 47
Flip-flop, J-K, 239
Focus-control adjustment with free-
  running sweep, 44-46
Free-running sweeps, 41-46
  centering-control adjustment, 43-44
  focus-control adjustment, 44-46
  intensity-control adjustment, 41-43
Frequency
  measurements, 36
  pulse, 16
  response, 36
Full adders and half adders, 236-237

G

Gain-control ranges and settings,
  55-57
Gaussian roll-off, 38
Glitches, 243
Growth and decay of waveform, 22-26

H

Half adders and full adders, 236-237
Hertz, 15
High-voltage
  capacitance-divider probe, 95-99
    inconsistent low-capacitance-
      probe response, 99
    stray-field interference, 97-99
  diode stack, 191-192
  ripple and filtering, 192-193
Horizontal
  and vertical linearity, 58-59
  oscillator and afc section, trouble-
    shooting, 165-180
    pattern-generator sync wave-
      forms, 175-177
    tube-type circuits, 177-180
  -sweep section, waveform tests in,
    181-198
    transistor horizontal-output ar-
      rangement, 184-193

Horizontal—cont
  -sweep section, waveform tests in
    troubleshooting
      SCR system, 181-184
      tube-type circuitry, 193-198
Horizontal Gain control, 53

I

Ideal pulses, 15
If section, signal tracing in, 111-127
Integrated-circuit if systems, 138-139
Integrators, 25-26
Intelligent oscilloscopes, 39-40
Intensity-control
  adjustment with free-running
    sweeps, 41-43
  setting vs. pattern size, 53-54
Intercarrier-sound
  signal processing, principles,
    217-225
  systems, tube-type, troubleshooting,
    225-228
  oscilloscope check for linearity,
    227-228
  phase-shift tests, 228
Intermittent monitoring, 125
Intermittents, if section, 123-127
Inverter, sync, operation of, 150
Invisible leading edge of pulse, 21
Isolating probe, 95

J

J-K flip-flop, 239

K

Kickback component, sync section,
  145

L

Linearity, horizontal and vertical,
  58-59
Lissajous family of waveforms, 32-35
Low
  -capacitance
    probe, 55-56
    probe response, inconsistent, 95
  -impedance demodulator probe,
    92-94

M

Measuring time intervals, 237-239
Medium-impedance demodulator
   probe, 92
Modified sine waveform, 29-30
Modular tv construction, note on,
   163-164
Modulation, amplitude, 35

N

NAND gate, 230
Natural frequency, 29
Negative
   -going sync pulse, ac type, 32
   picture, 122
Noise
   reduction, sync section, 145-150
   separator, 147
Nonlinearity, horizontal-amplifier, 58
NOR gate, 231-234

O

OR gate, 230-231
Oscilloscope
   dual-mode, 69
   how to operate, 41-80
     action of triggered-sweep con-
      trols, 69-80
     ac waveforms with dc compo-
      nents, 65-66
     basic scope with free-running
      sweeps, 41-46
      centering-control adjustment,
       43-44
      focus-control adjustment,
       44-46
      intensity-control adjustment,
       41-43
     dc voltage components and mea-
      surements, 63-65
     dual-trace display modes, 79-80
     peak-to-peak and peak voltage
      measurements, 60-63
     sync function, 66-68
     vertical and horizontal linearity,
      58-59
   intelligent, 39-40

Oscilloscope—cont
   introduction to, 11-40
     areas of application, two, 11
     basic waveforms, 13-35
      cornering of, 30
      growth and decay, 22-26
      hidden distortion, 19-21
      Lissajous family, 32-35
      modified sine, 29-30
      pulse polarities, 30-32
      steady state vs. transient, 26-29
     intelligent oscilloscopes, 39-40
     test classifications
      frequency measurements, 36
      frequency response, 36
      phase checks, 36
      signal tracing, 35
      time measurement, 36
      transient tests, 36
     time/frequency vs. data-domain,
      11-12
   probes, using, 81-104
     construction and operation of de-
      modulator probes, 88-95
      double-ended probe, 94-95
      low-impedance probe, 92-94
      medium-impedance probe, 92
      shunt-detector probe with RC
       filter, 90-91
     high-voltage capacitance-divider
      probe, 95-99
     inconsistent low-capacitance-
      probe response, 99
     stray-field interference, 97-99
     low-capacitance-probe construc-
      tion, 84-88
     overview of application in tv cir-
      cuitry, 99-101
     resistive "isolating" probe, 95
     special types of probes, 104
     vertical interval test signal,
      102-103
     why probes are needed, 82-84
   storage, 39-40
   techniques and pattern evaluation,
     sync section, 141-150
     kickback component, 145
     noise reduction, 145-150
     operation of sync inverter, 150

Oscilloscope—cont
  wideband, 38
Output stage operation, vertical-sweep
    section, 210-212
Overmodulation, amplitude, 35
Overshoot, 18

P

Pattern
  brightness, 54
  evaluation, sync section, 141-150
    kickback component, 145
    noise reduction, 145-150
    operation of sync inverter, 150
    -generator sync waveforms, 175-177
Peak
  voltage, sine wave, 13
  -to-peak
    and peak voltage measurements,
      60-63
    voltage, 15
Peaker circuit operation, vertical-
    sweep section, 209-210
Period, pulse, 16
Persistence of vision, 43
Phase
  checks, 36
  detector and color-sync action, 155
  -locked loop, 155
Picture, poor quality, 119-123
Pincushion
  correction, vertical-sweep section,
    206
  transformer, 189-191
Poor picture quality, 119-123
Position controls, 59
Positive-going sync pulse, ac type, 30
Power switch, turning on, 41
Preshoot, 18
Probe, low-capacitance, 55-56
Probes, oscilloscope, using, 81-104
  construction and operation of de-
      modulator probes, 88-95
    double-ended probe, 94-95
    low-impedance probe, 92-94
    medium-impedance probe, 92
    shunt-detector probe with RC
      filter, 90-91

Probes, oscilloscope, using—cont
  high-voltage capacitance-divider
      probe, 95-99
    inconsistent low-capacitance-
      probe response, 99
    stray-field interference, 97-99
  low-capacitance-probe construction,
    84-88
  overview of application in tv cir-
      cuitry, 99-101
  resistive "isolating" probe, 95
  special types of probes, 104
  vertical interval test signal, 102-103
  why probes are needed, 82-84
Propagation delay, 243
Pulse
  duty cycle, 16
  fall, 18
  frequency, 16
  ideal, 15
  period, 16
  polarities, 30-32
  repetition time, 16
  rise, 18
  waveshapes, digital, 241-247
  width, 16
Push-pull demodulator probe, 94-95

Q

Quality (Q) of circuit, 29

R

Race condition, 244
RAM, oscilloscope with, 12
Repetition time, pulse, 16
Resistive
  "isolating" probe, 95
  voltage division, probe, 86
Rf amplifiers, troubleshooting,
    105-111
Rise
  pulse, 18
  time, 18
Ringing, 19
Ripple-carry counter, 241
Rms voltage, sine wave, 13
Rounding, 18

S

Sag, distortion factor, 19
SCR system, horizontal-output, troubleshooting, 181-184
Settling time, 19
Shunt-detector probe with RC filter, 90-91
Signal tracing
  description, 35
  in rf, if, and video amplifiers, 105-139
    in if section, 111-127
    integrated-circuit if systems, 138-139
    in video amplifier, 127-130
    television station interference, 137-138
    transistor replacement, 139
    troubleshooting rf amplifiers, 105-111
  in sound-if and audio sections, 217-228
    troubleshooting tube-type intercarrier-sound systems, 225-228
      oscilloscope check for linear operation, 227-228
      phase-shift tests, 228
  in sync section, 141-164
    color-sync troubleshooting, 150-159
      dual-time-constant sync-clipper action, 155-159
      phase detector and color-sync action, 155
    oscilloscope techniques and pattern evaluation, 141-150
      kickback component, 145
      noise reduction, 145-150
      operation of sync inverter, 150
    sync troubleshooting in older receivers, 159-163
Sine-wave voltage
  average, 13
  peak, 13
  rms, 13
Slope and trigger level control, 69-74
Slope control, 69

Sound-if and audio sections, signal tracing in, 217-228
  principles of intercarrier-sound signal processing, 217-225
  troubleshooting audio section, 223-225
  troubleshooting tube-type intercarrier-sound systems, 225-228
    oscilloscope check for linear operation, 227-228
    phase-shift tests, 228
Spectrum analyzer, 36
Speed, sweep, and distortion, 19
Square-wave distortions, 130
Stability control, 77
Steady-state vs. transient waveforms, 26-29
Stepped corner, waveform, 30
Storage oscilloscope, 39-40
Stray-field interference, probe, 97-99
Stripping action, sync section, 145
Sweep speed and distortion, 19
Sync
  compression, 131
  -control adjustment, 47-50
  function, 66-68
  section, signal tracing in, 141-164
    color-sync troubleshooting, 150-159
      dual-time-constant sync-clipper action, 155-159
      phase detector and color-sync action, 155
    oscilloscope techniques and pattern evaluation, 141-150
      kickback component, 145
      noise reduction, 145-150
      operation of sync inverter, 150
    sync troubleshooting in older receivers, 159-163
  troubleshooting in older receivers, 159-163
Sync Amplitude control, 47-50
Synchronous binary counter, 241

T

Television station interference, 137-138

Terminal facilities and control functions, *see* Control functions
Tests classifications, oscilloscope frequency
  measurements, 36
  response, 36
  phase checks, 36
  signal tracing, 35
  time measurement, 36
  transient tests, 36
Tests, waveform, in horizontal-sweep section, 181-198
  transistor horizontal-output arrangement, 184-193
    high-voltage ripple and filtering, 192-193
    with
      high-voltage diode stack, 191-192
      pincushion transformer, 189-191
    troubleshooting
      SCR system, 181-184
      tube-type circuitry, 193-198
Time
  -base controls, 74-77
  intervals, measuring, 237-239
  measurement, 36
Time Base controls, 74
Timing diagram, 234
Toggle mode, 241
Tolerance, waveshape, 19
Transient tests, 36
Transistor
  horizontal-output arrangement, 184-193
    high-voltage ripple and filtering, 192-193
  replacement, 139
Traveling detectors, 92
Trigger
  controls, 77-79
  level and slope control, 69-74
Triggered-sweep controls, action of, 69-80
  time-base controls, 74-77
  trigger
    controls, 77-79
    level and slope controls, 69-74

Trigger Level control, 69
Troubleshooting
  afc and horizontal-oscillator section, 165-180
  audio section, tv, 223-225
  digital-logic, 229-247
    data-domain displays, 247
    operating waveforms, 235-247
    waveform relationships, 229-235
  SCR system, horizontal-output, 181-184
  tube-type
    horizontal-output circuitry, 193-198
    intercarrier-sound systems, 225-228
  vertical-sweep section, 199-215
    principles of operation, 199-212
      blocking oscillator arrangement, 207-208
      output stage operation, 210-212
      peaker circuit operation, 209-210
      pincushion correction, 206
      vertical blanking, 203
    tube-type circuitry, 212-215
      coupling-capacitor checks, 214
      feedback loop, 214-215

V

Variable Time CM control, 74
Vectorgram, 11
Vectorscope test, 32
Vernier sweep-frequency control, 47
Vertical
  and horizontal linearity, 58-59
  blanking, 203
  interval test signal, 102-103
  -sweep section, troubleshooting, 199-215
    principles of operation, 199-212
    tube-type circuitry, 212-215
Vertical Gain control, 53
Vertical Position control, 43
Video amplifier, signal tracing in, 127-130
VITS, 101-103

Voltage, sine-wave
    average, 13
    peak, 13
    rms, 13

W

Waveform
    checks, 36
    compression and clipping, 131
    tests in horizontal-sweep section,
        181-198
        transistor horizontal-output ar-
            rangement, 184-193
        troubleshooting tube-type cir-
            cuitry, 193-198
Waveforms
    basic, 13-35
        cornering of, 30
        growth and decay, 22-26
        hidden distortion, 19-21

Waveforms—cont
    basic
        Lissajous family, 32-35
        modified sine, 29-30
        pulse polarities, 30-32
        steady-state vs. transient, 26-29
    operating, digital-logic, 235-247
        binary counters, 239-241
        digital pulse waveshapes, 241-247
        half adders and full adders,
            236-237
        measuring time intervals, 237-239
    relationships, digital-logic, 229-235
Weak picture, 118
White compression, 131
Wideband oscilloscope for color tv, 38
Width, pulse, 16

X

Xor gate, 236-237